Collins

2014 GUIDE
to the
NIGHT SKY

Storm Dunlop and Wil Tirion

Published by Collins
An imprint of HarperCollins Publishers
Westerhill Road
Bishopbriggs
Glasgow G64 2QT

In association with
Royal Museums Greenwich,
the group name for the National Maritime Museum,
Royal Observatory Greenwich,
Queen's House and *Cutty Sark* 2013
www.rmg.co.uk

© Storm Dunlop and Wil Tirion
First published 2013
ISBN 978-0-00-754074-7
Imp 001

The contents of this publication are believed correct at the time of printing.
Nevertheless the publisher can accept no responsibility for errors or omissions,
changes in the detail given or for any expense or loss thereby caused.

British Library Cataloguing in Publication Data
A catalogue record for this book is available from the British Library

Printed in China by South China Printing Co. Ltd

If you would like to comment on any aspect of this book, please write to
Collins Maps, HarperCollins Publishers, Westerhill Road, Bishopbriggs, Glasgow G64 2QT
email: collinsmaps@harpercollins.co.uk
or visit our website at: www.collinsmaps.com

Contents

Month-by-month guide

Introduction

The aim of this Guide is to help people find their way around the night sky, by showing how the stars that are visible change from month to month and by including details of various events that occur throughout the year. The objects and events described may be observed with the naked eye, or nothing more complicated than a pair of binoculars.

The conditions for observing naturally vary over the course of the year. During the summer, twilight may persist throughout the night and make it difficult to see the faintest stars. There are three recognized stages of twilight: civil twilight, when the Sun is less than 6° below the horizon; nautical twilight, when the Sun is between 6° and 12° below the horizon; and astronomical twilight when the Sun is between 12° and 18° below the horizon. Full darkness occurs only when the Sun is more than 18° below the horizon. During nautical twilight, only the very brightest stars are visible. During astronomical twilight, the faintest stars visible to the naked eye may be seen directly overhead, but are lost at lower altitudes. As the diagram shows, during the summer months full darkness never occurs at the latitude of London, and at Edinburgh nautical twilight persists throughout the whole

night, so at that latitude only the very brightest stars are visible.

Another factor that affects the visibility of objects is the amount of moonlight in the sky. At Full Moon, it may be very difficult to see some of the fainter stars and objects, and even when the Moon is at a smaller phase it may seriously interfere with visibility if it is near the stars or planets in which you are interested. A full lunar calendar is given for each month and may be used to see when nights are likely to be darkest and best for observation.

The celestial sphere

All the objects in the sky (including the Sun, Moon, and stars) appear to lie at some indeterminate distance on a large sphere, centred on the Earth. This *celestial sphere* has various reference points and features that are related to those of the Earth. If the Earth's rotational axis is extended, for example, it points to the North and South Celestial Poles, which are thus in line with the North and South Poles on Earth. Similarly, the *celestial equator* lies in the same plane as the Earth's equator, and divides the sky into northern and southern hemispheres. Because this Guide is written for use in Britain and Ireland, the area

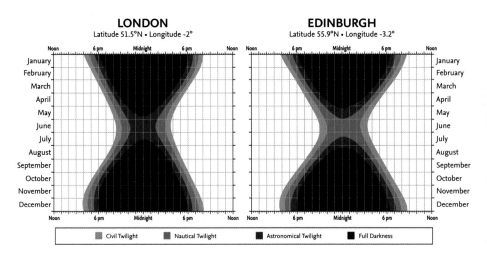

| LONDON | EDINBURGH |
| Latitude 51.5°N • Longitude -2° | Latitude 55.9°N • Longitude -3.2° |

Civil Twilight Nautical Twilight Astronomical Twilight Full Darkness of the sky that it describes includes the whole of the northern celestial hemisphere and those portions of the southern that become visible at different times of the year. Stars in the far south, however, remain invisible throughout the year, and are not included.

It is useful to know some of the special terms for various parts of the sky. As seen by an observer, half of the celestial sphere is invisible, below the horizon. The point directly overhead is known as the **zenith**, and the (invisible) one below one's feet as the **nadir**. The line running from the north point on the horizon, up through the zenith and then down to the south point is the **meridian**. This is an important invisible line in the sky, because objects are highest in the sky, and thus easiest to see, when they cross the meridian in the south. Objects are said to **transit**, when they cross this line in the sky.

In this book, reference is frequently made in the text and in the diagrams to the standard compass points around the horizon. The position of any object in the sky may be described by its **altitude** (measured in degrees above the horizon), and its **azimuth** (measured in degrees from north, 0°, through east, 90°, south, 180°, and west, 270°). Experienced amateurs and professional astronomers also use another system of specifying locations on

the celestial sphere, but that need not concern us here, where the simpler method will suffice.

The celestial sphere appears to rotate about an invisible axis, running between the north and south celestial poles. The location (i.e., the altitude) of the celestial poles depends entirely on the observer's position on Earth or, more specifically, their latitude. The charts in this book are produced for the latitude of 50°N, so the North Celestial Pole (NCP) is 50° above the northern horizon. The fact that the NCP is fixed relative to the horizon means that all the stars within 50° of the pole are always above the horizon and may, therefore, always be seen at night, regardless of the time of year. The northern circumpolar region is an ideal place to begin learning the sky, and ways to identify the circumpolar stars and constellations will be described shortly.

The ecliptic and the zodiac

Another important line on the celestial sphere is the Sun's apparent path against the background stars – in reality the result of the Earth's orbit around the Sun. This is known as the **ecliptic**. The point where the Sun, apparently moving along the ecliptic, crosses the celestial equator from south to north is known as the vernal (or spring) equinox, which occurs around March 21. At this time

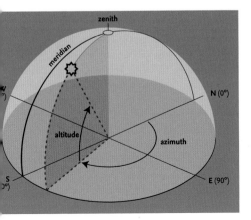

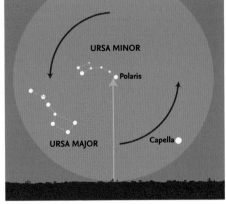

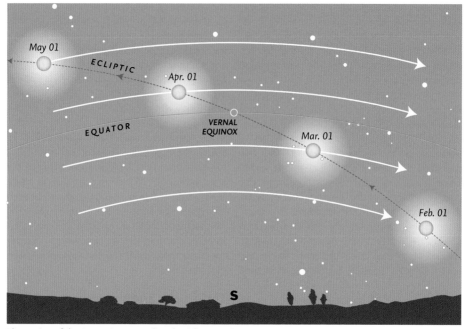

The motion of the Sun as it crosses the celestial equator in spring.

(and at the autumnal equinox, on September 22 or 23, when the Sun crosses the celestial equator from north to south) day and night are almost exactly equal in length. (There is a slight difference, but that need not concern us here.) The vernal equinox is currently located in the constellation of Pisces, and is important in astronomy because it defines the zero point for a system of celestial coordinates, which is, however, not used in this Guide.

The Moon and planets are to be found in a band of sky that extends 8° on either side of the ecliptic. This is because the orbits of the Moon and planets are inclined at various angles to the ecliptic (i.e., to the plane of the Earth's orbit). This band of sky is known as the Zodiac, and when originally devised, consisted of twelve **constellations**, all of which were considered to be exactly 30° wide. When the constellation boundaries were formally established by the International Astronomical

Union in 1933, the exact extent of most constellations was altered, and nowadays, the ecliptic passes through thirteen constellations. Because of the boundary changes, the Moon and planets may actually pass through several other constellations that are adjacent to the original twelve.

The constellations

Since ancient times, the celestial sphere has been divided into various constellations, most dating back to antiquity and usually associated with certain myths or legendary people and animals. Nowadays, the boundaries of the constellations have been fixed by international agreement and their names (in Latin) are largely derived from Greek or Roman originals. Some of the names of the most prominent stars are of Greek or Roman origin, but many are derived from Arabic names. Many bright stars have no individual names, and for many years, stars were identified by terms such as 'the star in Hercules' right foot'. A more sensible scheme was introduced by the German astronomer Johannes Bayer in the early 17th century. Following his scheme – which is still used today – most of the brightest stars are identified by a Greek letter followed by the genitive form of the constellation's Latin name. An example is the Pole Star, also known as Polaris and α Ursae Minoris. The Greek alphabet is shown on page 93, and also on page 93 is a list of all the constellations that may be seen from latitude 50°N, together with their genitive forms, and their English names. Other naming schemes exist for fainter stars, but are not used in this book.

Asterisms

Apart from the constellations (88 of which cover the whole sky), certain groups of stars, which may form a small part of a larger constellation, are readily recognizable and have been given individual names. These groups are known as *asterisms*, and the most famous (and well-known) is the 'Plough', the common name for the seven brightest stars in the constellation of Ursa Major, the Great Bear. The names and details of some asterisms mentioned in this book are given in the list on page 94.

Magnitudes

The brightness of a star, planet, or other body is frequently given in magnitudes (mag.). This is a mathematically defined scale where larger numbers indicate a fainter object. The scale extends beyond the zero point to negative numbers, for very bright objects. (Sirius, the brightest star in the sky is mag. -1.4) Most observers are able to see stars down to about mag. 6, under very clear skies.

The Moon

As it gradually passes across the sky from west to east in its orbit around the Earth, the Moon moves by approximately its diameter (about half a degree) in an hour. Normally, in its orbit around the Earth, the Moon passes above or below the direct line between Earth and Sun (at New Moon) or outside the area obscured by the Earth's shadow (at Full Moon). Occasionally, however, the three bodies are more-or-less perfectly aligned to give an eclipse: a solar eclipse at New Moon, or a lunar eclipse at Full Moon. Depending on the exact circumstances, a solar eclipse may be merely partial (when the Moon does not cover the whole of the Sun's disk); annular (when the Moon is too far from Earth in its orbit to appear large enough to hide the whole of the Sun); or total. Total and annular eclipses are visible from very restricted areas of the Earth, but partial eclipses are normally visible over a wider area.

Somewhat similarly, at a lunar eclipse, the Moon may pass through the outer zone of the Earth's shadow, the penumbra (in a penumbral eclipse, which is not generally perceptible to the naked eye); so that just part of the Moon is within the darkest part of the Earth's shadow, the umbra (in a partial eclipse); or completely within the umbra (in a total eclipse). Unlike solar eclipses, lunar eclipses are visible from large areas of the Earth.

Occasionally, as it moves across the sky, the Moon passes between the Earth and individual planets or distant stars giving rise to an *occultation*. As with solar eclipses, such occultations are visible from restricted areas of the world.

The Planets

Because the planets are always moving against the background stars, they are treated in some detail in the monthly pages and information is given when they are close to other planets, the Moon, or any of five bright stars that lie near the ecliptic. Such events are known as *appulses* or, more frequently, as *conjunctions*. (There are technical differences in the way these terms are defined – and should be used – in astronomy,

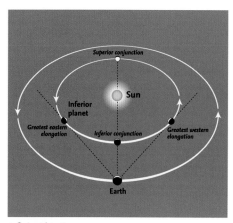

Inferior planet.

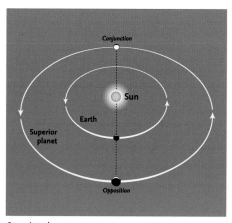

Superior planet.

but these need not concern us here.) The positions of the planets are shown for every month on a special chart of the ecliptic.

The term conjunction is also used when a planet is either directly behind or in front of the Sun, as seen from Earth. (Under normal circumstances it will then be invisible.) The conditions of most favourable visibility depend on whether the planet is one of the two known as *inferior planets* (Mercury and Venus) or one of the three *superior planets* (Mars, Jupiter and Saturn) that are covered in detail. (Details of the fainter superior planets, Uranus and Neptune, are not included in this Guide, although one special chart is given for Uranus on page 18.)

The inferior planets are most readily seen at eastern or western *elongation*, when their angular distance from the Sun is greatest. For superior planets, they are best seen at *opposition*, when they are directly opposite the Sun in the sky, and cross the meridian at local midnight.

It is often useful to be able to estimate angles on the sky, and approximate values may be obtained by holding one hand at arm's length. The various angles are shown in the diagram, together with the separations of the various stars in the Plough.

Meteors

At some time or other, nearly everyone has seen a *meteor* – a 'shooting star' – as it flashed across the sky. The particles that cause meteors – known technically as 'meteoroids' – range in size from that of grain of sand (or even smaller), to the size of a pea. On any night of the year there are occasional meteors, known as *sporadics*, that may travel in any direction. These occur at a rate that is normally between three and eight in an hour. Far more important, however, are *meteor showers*, which occur at fixed periods of the year, when the Earth encounters a trail of particles left behind by a comet or, very occasionally, by a minor planet (asteroid). Meteors always appear to diverge from a single point on the sky, known as the *radiant*, and the radiants of major showers are shown on the charts. Meteors that come from a circular area, 8° in diameter around the radiant are classed as belonging to the particular shower. All others that do not come from that area are sporadics (or, occasionally from another shower that is active at the same time). A list of the major meteor showers is given on page 17.

Although the positions of the various shower radiants are shown on the charts, looking directly at the radiant is not the most

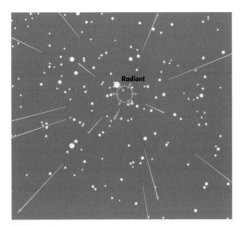

Meteor shower.

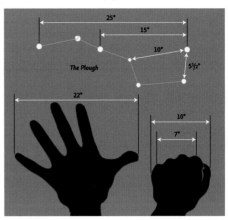

Measuring angles in the sky.

effective way of seeing meteors. They are most likely to be noticed if one is looking about 40–45° away from the radiant position. (This is approximately two hand-spans as shown in the diagram for measuring angles.)

Other objects

Certain other objects may be seen with the naked eye under good conditions. Some were given names in antiquity – Praesepe is one example – but many are known by what are called 'Messier numbers', the numbers in a catalogue of nebulous objects compiled by Charles Messier in the late 18th century. Some, such as the Andromeda Galaxy, M31, and the Orion Nebula, M42, may be seen by the naked eye, but all those given in the list will benefit from the use of binoculars.

Apart from galaxies, such as M31, which contain thousands of millions of stars, there are also two types of cluster: open clusters, such as M45, the Pleiades, which may consist of a few dozen to some hundreds of stars; and globular clusters, such as M13 in Hercules, which are spherical concentrations of many thousands of stars. One or two gaseous nebulae, consisting of gas illuminated by stars within them are also visible. The Orion Nebula, M42, is one, and is illuminated by the group

of four stars, known as the Trapezium, which may be seen within it by using a good pair of binoculars.

Some interesting objects

Messier number	Name Constellation	Type Maps (months)
—	**Hyades** Taurus	open cluster Sep. – Apr.
M8	**Lagoon Nebula** Sagittarius	gaseous nebula Jun. – Sep.
M11	**Wild Duck Cluster** Scutum	open cluster May – Oct.
M13	**Herculus Cluster** Hercules	globular cluster Feb. – Nov.
M15	— Pegasus	globular cluster Jun. – Dec.
M22	— Sagittarius	globular cluster Jun. – Sep.
M27	**Dumbbell Nebula** Vulpecula	planetary nebula May – Dec.
M31	**Andromeda Galaxy** Andromeda	galaxy All year
M35	— Gemini	open cluster Oct. – May.
M42	**Orion Nebula** Orion	gaseous nebula Nov. – Mar.
M44	**Praesepe** Cancer	open cluster Nov. – Jun.
M45	**Pleiades** Taurus	open cluster Aug. – Apr.

The Northern Constellations

The northern circumpolar stars

In learning to find one's way around the sky and to identify the constellations, the northern circumpolar stars are the key starting point for northern-hemisphere observers. Not only are they visible at any time of the year, but nearly everyone living in the northern hemisphere is familiar with the seven stars of the Plough – known as the Big Dipper in North America – an asterism that forms part of the large constellation of Ursa Major. This is where we start. There are five main constellations and one all-important star to identify:

- **Ursa Major** (The Great Bear)
- **Polaris**
- **Ursa Minor** (The Little Bear)
- **Cassiopeia**
- **Cepheus**
- **Draco** (The Dragon)

Ursa Major

Because of the movement of the stars caused by the passage of the seasons, **Ursa Major** lies in different parts of the evening sky at different periods of the year. The diagram shows its position for the four main seasons to give you

> Continued on page 12

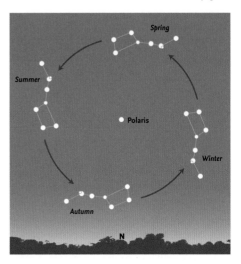

The trails of the stars around the north celestial pole, on a long exposure photograph.

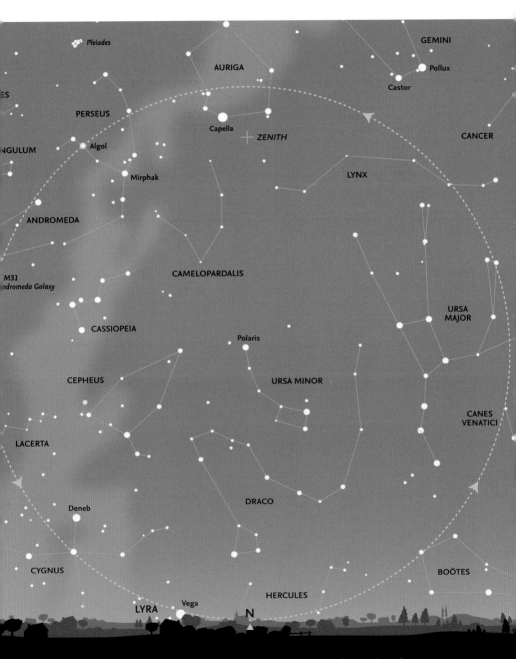

Pleiades

GEMINI

AURIGA

Pollux

Castor

PERSEUS

CANCER

ES

Capella
+ ZENITH

NGULUM · Algol

LYNX

Mirphak

ANDROMEDA

CAMELOPARDALIS

M31
ndromeda Galaxy

URSA
MAJOR

CASSIOPEIA

Polaris

CEPHEUS

URSA MINOR

CANES
VENATICI

LACERTA

DRACO

Deneb

CYGNUS

BOÖTES

HERCULES

LYRA   Vega   N

The stars and constellations inside the circle are always above the horizon, seen from our latitude.

THE NORTHERN CONSTELLATIONS   **11**

a guide as to where to look. The seven stars remain visible throughout the year anywhere north of latitude 40°N. Even at the latitude (50°N) for which the charts in this book are drawn, many of the stars in the southern portion of the constellation are hidden below the horizon for part of the year.

Ursa Major is a very large constellation, but initially few people are familiar with the extended groups of stars that form part of it and lie well to the south and east. In the centre of the curve of stars that form the 'tail' of the Great Bear (or the 'handle' of the Dipper) lies the small constellation of Canes Venatici, which consists of two moderately bright stars and a scattering of fainter ones.

Polaris

Once you have identified the seven bright stars of the Plough, locate the two stars, α and β UMa farthest away from the 'tail'. These two, named **Dubhe** and **Merak**, respectively, are known as the 'Pointers'. A line from Merak (the southernmost) to Dubhe, extended to about five times their separation, leads more-or-less directly to a fairly isolated bright star. This is the Pole Star, *Polaris*, or α Ursae Minoris and the brightest star in the constellation. All the stars in the northern sky appear to rotate around it. In fact, it lies slightly less than one degree away from the true pole, and a trailed photograph of the northern sky shows that it traces a tiny circle (about 1.5 degrees across) around the pole.

Apart from this small variation, the elevation of Polaris above the northern horizon is always equal to the observer's latitude. For sailors, this was an extremely useful property, and was used as a guide to navigation by generations of sailors from the time of the

Finding Polaris and Ursa Minor.

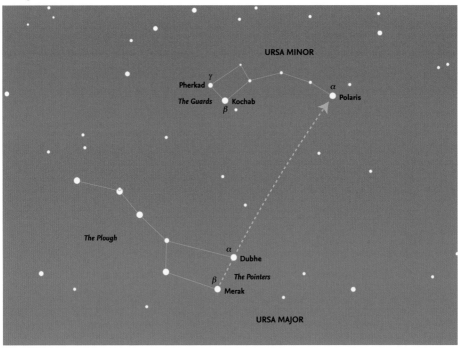

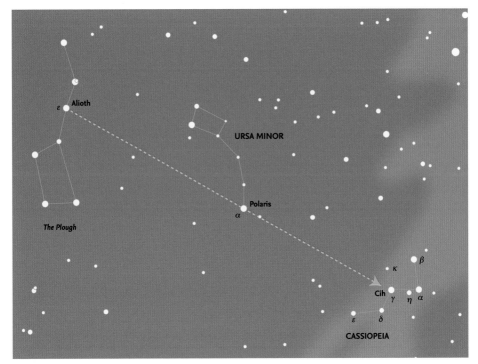

Finding Cassiopeia from the Plough and Polaris.

Greeks to comparatively modern times. Now it is our key to learning the sky.

Ursa Minor

In contrast to Ursa Major, **Ursa Minor** is a small constellation, consisting of little more than the relatively faint stars that form the 'Little Dipper' (as it is known in North America). Strangely, the asterism has no common name in Britain, other than the Little Bear, despite being superficially similar to the seven stars of the Plough. Just five stars – including a moderately close pair – form the body of the constellation, with another three forming the 'tail' at the tip of which lies Polaris – the end of the 'handle' of the Little Dipper. The two stars farthest from the pole, **Kochab** and **Pherkad** (β and γ Ursae Minoris, respectively) are known in English-speaking countries as 'The Guards'.

Cassiopeia

On the opposite side of Polaris and the North Pole from Ursa Major lies **Cassiopeia**. It has a highly distinctive shape, appearing as five stars that form a letter 'W' or 'M' depending on its orientation. Provided the sky is reasonably clear of clouds, you will nearly always be able to see either Ursa Major or Cassiopeia, and thus be able to orientate yourself on the sky.

To find Cassiopeia from Ursa Major, start with **Alioth** (ε Ursae Majoris), the first star in the tail of the Bear. A line from this star extended through Polaris points directly towards **Cih** (γ Cassiopeiae), the central star of the five. Cassiopeia lies right in the centre of the band of the Milky Way, so on a clear night a large number of fainter stars surround the five main stars, and this may sometimes make identification slightly more difficult for

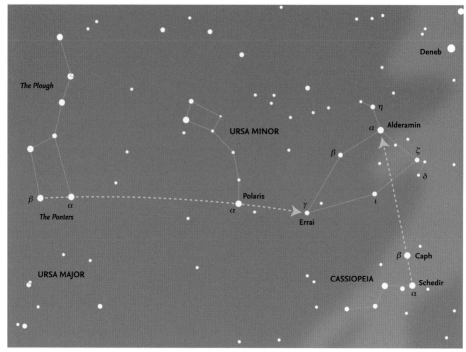

Locating the constellation of Cepheus.

beginners. Even when the Milky Way is not readily visible because of moonlight or light pollution, some people are still occasionally confused by the presence of the moderately bright stars, η and κ Cassiopeia that lie nearby. With just a little practice, however, Cassiopeia becomes easy to recognize.

Cepheus

Although the constellation of **Cepheus** is circumpolar, it is not nearly as well known as Ursa Major, Ursa Minor or Cassiopeia. This is partly because it does not form a highly distinctive pattern on the sky, and also because some of its stars are faint. Its shape is often likened to (and does actually resemble) the gable end of a house, with the 'ground' lying in the Milky Way not far from Cassiopeia, and the tip of the gable pointing towards the pole. Its brightest star, **Alderamin** (α Cephei), lies in

the Milky-Way region, at the 'bottom right-hand corner' of the figure. A line from the stars **Schedir** and **Caph** (α and β Cassiopeiae), extended about three times their separation, passes just north of Alderamin. The star at the tip of the 'gable', **Errai** (γ Cephei) lies close to the line from Polaris (α UMa) to Caph (β Cas). The line from the Pointers to Polaris, if extended, also passes just 'below' the tip of the 'gable'.

Draco

The last of the circumpolar constellations that is fairly easy to identify is **Draco**, although it is such a long, winding constellation that it takes some experience to recognize it easily. It consists of a quadrilateral of stars, known, logically enough, as the 'Head of Draco' (and also the 'Lozenge'), and a long chain of stars forming the neck and body of the dragon.

Finding the Head of Draco is not particularly easy using just circumpolar stars. Locate the two stars Phad and Megrez (γ and δ Ursae Majoris) at the opposite end of the bowl of the Plough from the Pointers. Extend a line from Phad (γ UMa) through Megrez (δ UMa) by about eight times their separation. This takes you right across the sky below the Guards in Ursa Minor, and close to **Grumium** (ξ Draconis) at one corner of the quadrilateral. The brightest star in the constellation, **Etamin** (γ Draconis) lies farther to the south. From the head of Draco, the constellation first runs northeast to δ and ε Dra, then doubles back southwards, before winding its way round between Ursa Minor and Ursa Major, through **Thuban** (α Draconis), before ending at **Giausar**

(λ Draconis) almost on the line between the Pointers and Polaris.

The circumpolar chart shows that there is a large area of sky between Ursa Major and Cassiopeia that has relatively few bright stars. There is one constellation here, **Camelopardalis**, which is always circumpolar, but which is so faint that it is easier to learn to find it after you have gained some experience, and are familiar with some of the constellations (Perseus and Auriga) that lie farther south. Another dim constellation, **Lynx** – a long, straggling line of faint stars – is also rather difficult to see. Large parts of the constellations of Auriga, Cygnus, and Perseus are often circumpolar, depending on your exact latitude.

Finding the head of Draco.

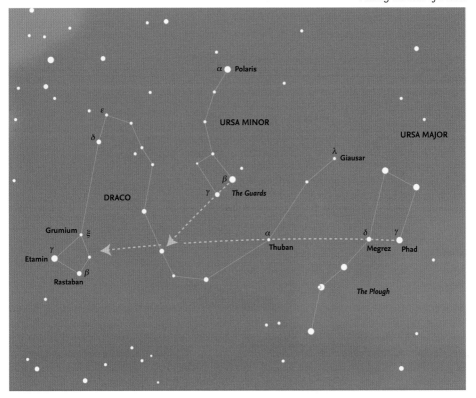

Introduction to the month-by-month guide

The monthly charts

The pages devoted to each month contain a pair of charts showing the appearance of the night sky, looking north and looking south. The charts (as with all the charts in this book) are drawn for the latitude of 50°N, so observers farther north will see slightly more of the sky on the northern horizon, and slightly less on the southern. These areas are, of course, those most likely to be affected by poor observing conditions caused by haze, mist or smoke. In addition, stars close to the horizon are always dimmed by atmospheric absorption, so sometime the faintest stars marked on the charts may not be visible.

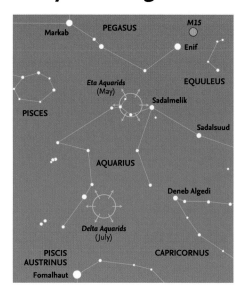

The three times shown for each chart require a little explanation. The charts are drawn to show the appearance at 23:00 GMT for the 1st of each month. The same appearance will apply an hour earlier (22:00 GMT) on the 15th, and yet another hour earlier (21:00 GMT) at the end of the month (shown as the 1st of the following month). GMT is identical to the Universal Time (UT), used by astronomers around the world. In Europe, Summer Time is introduced in March, so the March charts apply to 23:00 GMT on March 1, 22:00 GMT on March 15, but 22:00 BST (British Summer Time) on April 1. The change back from Summer Time (in Europe) occurs in October, so the charts for that month apply to 00:00 BST for October 1, 23:00 BST for October 15, and 21:00 GMT for November 1.

The charts may be used for earlier or later times during the night. To observe two hours earlier, use the charts for the preceding month; for two hours later, the charts for the next month.

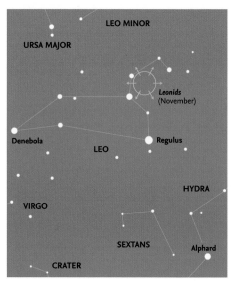

Meteors

The details of specific meteor showers are given in the months in which they come to maximum, regardless of whether they begin or end in other months. It should be noted that not all the respective radiants will be marked on the charts for that particular month, because the radiants may be below the horizon, or lie in constellations that are not readily visible during the month of maximum. For this reason, special charts for the Eta and Delta Aquarids (May and July, respectively)

Shower	Dates of activity	Date of maximum	Possible hourly rate
Quadrantids	January 1–6	January 3	70
Lyrids	April 19–25	April 21	10–15
Eta Aquarids	April 24 to May 20	May 5	35
Delta Aquarids	July 15 to August 20	July 29	20–25
Perseids	July 15 to August 20	August 12	80
Alpha Aurigids	August to October	August 28 & September 15	10
Orionids	October 15 to November 2	October 21	30
Taurids	October 15 to November 25	November 3	10
Leonids	November 15–20	November 17	around 60
Geminids	December 7–15	December 13	around 100

and the Leonids (November) are given here. As explained earlier, however, meteors from such showers may still be seen, because the most effective region for seeing meteors is some 40–45° away from the radiant, and that area of sky may well be above the horizon. For convenience, a table of all the meteor showers visible during the year is also given here.

The photographs

As an aid to identification – especially as some people find it difficult to relate charts to the actual stars they see in the sky – one or more photographs of constellations visible in any particular month are included. It should be noted, however, that because of the limitations of the photographic and printing processes, and the differences between the sensitivity of different individuals to faint starlight (especially in their ability to detect different colours), and the degree to which they have become adapted to the dark, the apparent brightness of stars in the photographs will not necessarily precisely match that seen by any one observer.

The Moon calendar

The Moon calendar is largely self-explanatory. It shows the phase of the Moon for every day of the month, with the exact times (in Unversal Time) of New Moon, First Quarter, Full Moon, and Last Quarter. Because the times are calculated from the Moon's actual orbital parameters, some of the times shown will, naturally, fall during daylight, but any difference is too small to affect the appearance of the Moon on that date.

The Moon

The section on the Moon includes details of any lunar or solar eclipses that may occur during the month (visible from anywhere on Earth). Similar information is given about any important occultations. Mainly, however, this section summarizes when the Moon passes close to planets or the five prominent stars close to the ecliptic. The dates when the Moon is closest to the Earth (at perigee) and farthest from it (at apogee) are shown in the monthly calendars, and only mentioned here when they are particularly significant, such as the nearest and farthest during the year.

The Planets

Brief details are given of the location, movement and brightness of the planets from Mercury to Saturn throughout the month. None of the planets can, of course, be seen when they are close to the Sun, so such periods are generally noted. All of the planets may sometimes lie on the opposite side of the Sun to the Earth (at what is known as superior conjunction), but in the case of the inferior

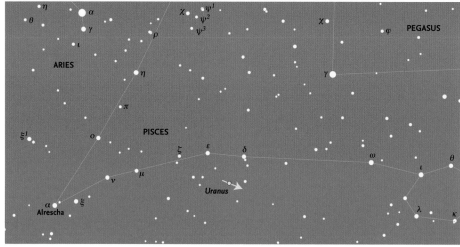

The position of Uranus in October 2014.

planets, Mercury and Venus they may also pass between the Earth and the Sun (at inferior conjunction) and are normally invisible for a longer or shorter period of time. Those two planets are normally easiest to see around either eastern or western elongation, in the evening or morning sky, respectively. Not every such elongation is favourable, however, so although every elongation is listed, only those where observing conditions are most favourable are shown in the individual diagrams of events.

The dates at which the superior planets reverse their motion (from direct motion to retrograde, and retrograde to direct) and of opposition (when a planet generally reaches its maximum brightness) are given. Some planets, especially distant Saturn, may spend most or all of the year in a single constellation. Jupiter and Saturn are normally easiest to see around opposition which occurs every year. Mars, by contrast, moves relatively rapidly against the background stars and in some years never comes to opposition.

Uranus is not included in the monthly details because it is generally too faint to be readily visible, although around opposition it becomes bright enough to be detectable in binoculars, or even with the naked eye under exceptionally dark skies. In 2014 it comes to opposition in the constellation of Pisces and the region is shown on the special diagram here. Unfortunately, in 2014 the nearly Full Moon is nearby on the date of actual opposition (October 7), which will cause interference for a couple of days. The planet should, however, be readily detectable for some days on either side of that date.

The ecliptic charts

Although the ecliptic charts are primarily designed to show the positions and motions of the major planets, they also show the motion of the Sun during the month. The light-tinted area shows the area of the sky that is invisible during daylight, but the darker area gives an indication of which constellations are likely to be visible at some time of the night. The closer a planet is to the border between dark and light, the more difficult it will be to see in the twilight. (If it is in the light area, of course, it will be invisible, drowned by the light from the Sun.)

The monthly calendar

For each month, a calendar shows details of significant events, including when planets are close to one another in the sky, close to the Moon, or close to any one of five bright stars that are spaced along the ecliptic. The times shown are given in Universal Time (UT), always used by astronomers throughout the year, and which is identical to Greenwich Mean Time (GMT). So, even during the summer months, they do not show Summer Time, which will always be one hour later than the time shown.

The diagrams of interesting events

Each month, a number of diagrams show the appearance of the sky when certain events take place. However, the exact positions of celestial objects and their separations greatly depend on the observer's position on Earth. When the Moon is one of the objects involved, because it is relatively close to Earth, there may be very significant changes from one location to another. Close approaches between planets or between a planet and a star are less affected by changes of location, which may thus be ignored.

The diagrams showing the appearance of the sky are drawn for the latitude of London, so will be approximately correct for most of Britain and Europe. However, for an observer farther north (say Edinburgh), a planet or star listed as being north of the

Moon will appear even farther north, whereas one south of the Moon will appear closer to it – or may even be hidden (occulted) by it. For an observer farther south than London, there will be corresponding changes in the opposite direction: for a star or planet south of the Moon the separation will increase, and for one north of the Moon the separation will decrease. For example, in February 2014 the calendar shows that Venus will be south of the Moon on February 26, but in the illustration, created for London, it is north of the Moon.

Ideally, details should be calculated for each individual observer, but this is obviously impractical. In fact, positions and separations are actually calculated for a theoretical observer located at the centre of the Earth.

So the details given regarding the positions of the various bodies should be used as a guide to their location. A similar situation arises with the times that are shown. These are calculated according to certain technical criteria, which need not concern us here. However, they do not necessarily indicate the exact time when two bodies are closest together. Similarly, dates and times are given, even if they fall in daylight, when the objects are likely to be completely invisible. However, such times do give an indication that the objects concerned will be in the same general area of the sky during both the preceding, and the following nights.

Key to the symbols used on the monthy star maps.

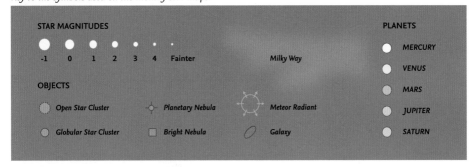

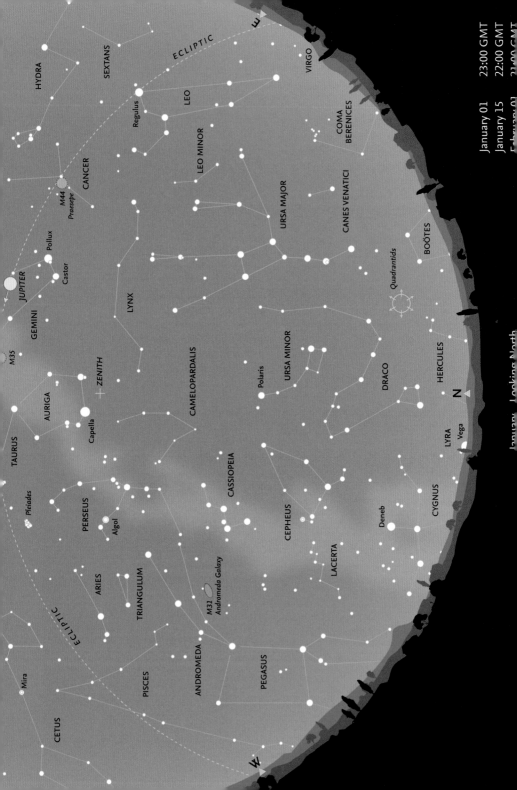

January 01   23:00 GMT
January 15   22:00 GMT
February 01   21:00 GMT

January - Looking North

N

W

E

ECLIPTIC

ECLIPTIC

ZENITH

HYDRA

SEXTANS

VIRGO

LEO

Regulus

COMA
BERENICES

LEO MINOR

CANCER

M44
Praesepe

URSA MAJOR

CANES VENATICI

Pollux

Castor

BOÖTES

JUPITER

Quadrantids

GEMINI

LYNX

URSA MINOR

M35

Polaris

HERCULES

AURIGA

CAMELOPARDALIS

DRACO

TAURUS

Capella

LYRA

Vega

CASSIOPEIA

CEPHEUS

CYGNUS

Pleiades

PERSEUS

Deneb

Algol

LACERTA

ARIES

TRIANGULUM

M31
Andromeda Galaxy

Mira

PISCES

ANDROMEDA

PEGASUS

CETUS

January – Looking North

Most of the important circumpolar constellations are easy to see in the northern sky at this time of year. **Ursa Major** stands more-or-less vertically above the horizon in the northeast, with the zodiacal constellation of **Leo** rising in the east. To the north, the stars of **Ursa Minor** lie below Polaris (the Pole Star). The head of **Draco** is low on the northern horizon, but may be difficult to see unless observing conditions are good. Both **Cepheus** and **Cassiopeia** are easily seen in the northwest, and even the faint constellation of **Camelopardalis** is high enough in the sky for it to be readily visible. **Auriga**, with bright **Capella** is high overhead, near the zenith, while **Perseus** and **Andromeda** are visible farther down towards the west, where the Great Square of **Pegasus** is approaching the horizon.

Meteors

There is one fairly strong meteor shower in January: the **Quadrantids**, which are visible on January 1–6, with the maximum on January 3. They are bright, bluish and yellowish-white meteors and because New Moon is on January 1 should be readily visible. At the maximum they may reach a rate of 70 meteors per hour.

They are named after the former constellation of **Quadrans Muralis** (the Mural Quadrant), an early form of astronomical instrument. The Quadrantid meteor radiant is now within the northernmost part of **Boötes**, roughly half-way between θ Boötis and τ Herculis.

Comet

On 8 January 2014, Comet **ISON** will appear close to Polaris, and on January 14–15 the Earth will pass close to its orbit, when its dust may cause a meteor shower, and possibly act as nuclei for noctilucent-cloud particles (p. 51). It will have passed perihelion (closest point to the Sun) on 28 November 2013, at a distance of 1,100,000 km, (0.983 AU)

but – as with most comets – its brightness is difficult to predict. If it survives perihelion passage (many comets break up close to the Sun), it may be visible to the naked eye in January.

The constellation of Orion dominates the sky during this period of the year, and is a useful starting point for recognizing other constellations in the southern sky. This photograph was taken with a 25-second exposure on an ordinary camera. Orion can be found in the southern part of the sky (see next page).

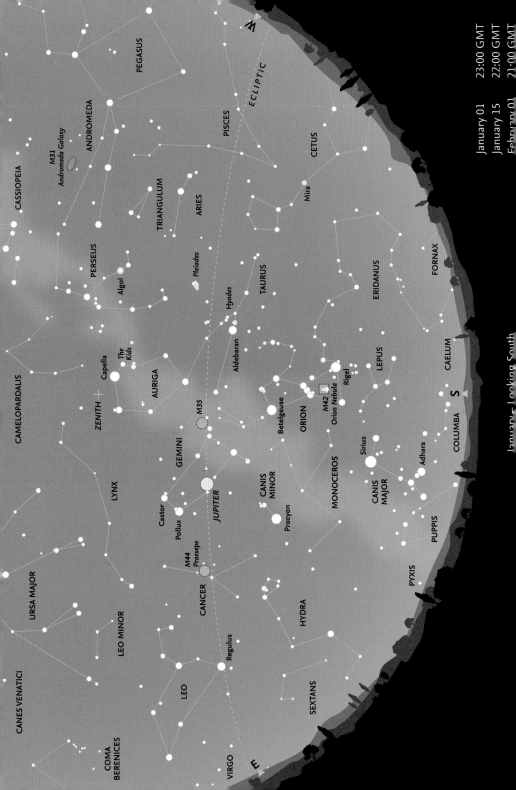

CANES VENATICI

COMA BERENICES

URSA MAJOR

LEO MINOR

LEO

CAMELOPARDALIS

LYNX

CASSIOPEIA

M31
Andromeda Galaxy

ANDROMEDA

PEGASUS

PERSEUS

Algol

TRIANGULUM

ARIES

PISCES

ECLIPTIC

The Kids

Capella

Pleiades

Hyades

TAURUS

CETUS

Mira

GEMINI

ZENITH +

AURIGA

Aldebaran

Castor

Pollux

M35

Betelgeuse

ORION

M42
Orion Nebula

Rigel

ERIDANUS

FORNAX

M44
Praesepe

CANCER

JUPITER

CANIS
MINOR

Procyon

MONOCEROS

Sirius

CANIS
MAJOR

LEPUS

CAELUM

S

COLUMBA

VIRGO

E

SEXTANS

HYDRA

PYXIS

PUPPIS

Adhara

January — Looking South

January 01   23:00 GMT
January 15   22:00 GMT
February 01   21:00 GMT

January – Looking South

At this time of year the southern sky is dominated by **Orion**. This is the most prominent constellation during the winter months, when it is visible at some time during the night. It has a highly distinctive shape, with a line of three stars that form the 'Belt'. To most observers, the bright star at the north-western corner of the constellation, **Betelgeuse**, (α Orionis), shows a reddish tinge, in contrast to the brilliant bluish-white colour of the bright star at the south-western corner, **Rigel**, (β Orionis). The three stars of the belt lie across the celestial equator. A vertical line of three 'stars' forms the 'Sword' that hangs to the south of the Belt. With good viewing conditions, the central 'star' appears as a hazy spot, even to the naked eye. This is actually the great Orion Nebula.

The line of Orion's Belt points up to the north-east towards **Taurus** (the Bull) and orange-tinted **Aldebaran**, (α Tauri). Close to Aldebaran, there is a conspicuous 'V' of stars, pointing down to the south-west, an open cluster called the **Hyades**. Farther along, the same line from Orion takes you close to a bright cluster of stars, the **Pleiades**, or Seven Sisters. Even the smallest pair of binoculars reveals this cluster as a beautiful group of bluish-white stars. The two most conspicuous of the other stars in Taurus lie directly above Orion, and form an elongated triangle with Aldebaran. The northernmost, β Tauri, was once considered to be part of the constellation of Auriga.

Also above Orion at this time of year is the constellation of **Auriga** (the Charioteer), with brilliant **Capella**, (α Aurigae), close to the zenith, directly overhead. Slightly to the west of Capella lies a small triangle of fainter stars, known as '**The Kids**'. (Ancient mythological representations of Auriga show him carrying two young goats.) Together with that northernmost bright star in Taurus, β Tauri, the body of Auriga forms a large pentagon on the sky, with the Kids lying part-way along the eastern side.

The Moon's phases for January

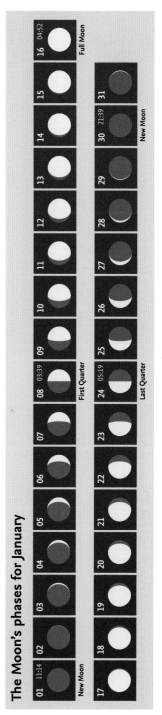

01 11:14 New Moon	**02**	...

There are two New Moons in January 2014, on the 1st and 30th of the month. The Full Moon on January 16 occurs when the Moon is at perigee (closest to the Earth) and is actually slightly larger in the sky, although the increase is not readily visible to the naked eye.

January – Moon and Planets

The Earth

The Earth reaches perihelion (the closest point to the Sun in its annual orbit) on 4 January 2014, at almost exactly noon (at 11:59 Universal Time, to be precise). Its distance is then 147,104,780 km (0.983 AU).

The Moon

The Moon passes close to several planets during the month. On January 2, the very young Moon (just one day old) is near **Venus** in the western sky. On January 15 (one day before Full Moon) it lies south of Jupiter, and later in the month (January 23–26) it passes **Mars**, **Spica** (α Virginis), and **Saturn** in the early-morning sky. At the end of the month (January 31 to February 1) the one- and two-day-old Moon is near **Mercury** in the west shortly after sunset.

The Planets

Jupiter is conspicuous in **Gemini** throughout the night. On January 5 it is at opposition at magnitude -2.2. On January 15 it is close to the nearly Full Moon. (It is closest at 06:08 on January 15, in the daylight sky.) **Venus** passes behind the Sun on January 10, and is thus invisible during the month. Early in the month, **Mars** rises shortly after midnight and **Saturn** about 04:00. At the end of the month both planets (in Virgo and Libra, respectively) rise about two hours earlier. It may be possible to glimpse **Mercury** in the evening twilight around greatest eastern elongation on January 31, close to the day-old Moon. There is an occultation of Saturn on January 25, visible from the southern hemisphere.

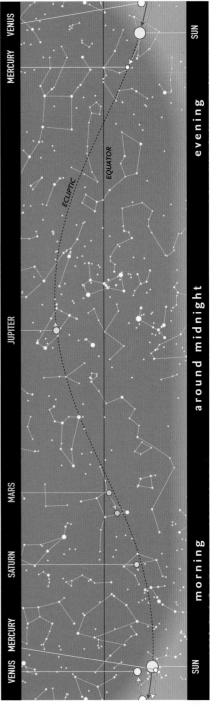

The path of the Sun and the planets along the ecliptic in January.

Calendar for January

01	11:14	New Moon
01	20:59	Moon at perigee
01	15:41	Mercury 6.6°S of Moon
01–06		Quadrantid meteor shower
02	11:46	Venus 2.0°S of Moon
03		Quadrantid meteor maximum
04	11:59	Earth at perihelion (147,104,780 km = 0.983 AU)
05	21:11	Jupiter at opposition (mag. -2.2)
07	10:00*	Mercury 6.4°S of Venus
08	03:39	First Quarter
10	12:24	Venus inferior conjunction
12	09:00	Aldebaran 2.6°S of Moon
15	06:08	Jupiter 4.9°N of Moon
16	03:58	Pollux 11.8°N of Moon
16	01:53	Moon at apogee
16	04:52	Full Moon
19	05:07	Regulus 5.2°N of Moon
23	06:29	Mars 3.7°N of Moon
23	09:44	Spica 1.3°S of Moon
24	05:19	Last Quarter
25	13:58	Saturn 0.6°N of Moon
25		Saturn occulted by Moon (N. Zealand, S. Pacific, S. South America, part of Antarctica
26	18:11	Antares 7.6°S of Moon
28	20:00*	Spica 4.9°S of Mars
29	02:36	Venus 2.3°N of Moon
30	21:39	New Moon
30	09:59	Moon at perigee
31	09:58	Mercury greatest elongation (18°E, mag. -0.7)

*Entries with an * are at this distance for an extended period around this time.*

Evening 17:00

10° *5°*

Moon

Venus

SW

January 2 • *The Moon and Venus in the evening sky. The actual conjunction occurs in daylight (11:46).*

Morning 5:00

10°

Castor
Pollux

Jupiter

Moon

Procyon

W
WNW

January 15 • *The Moon and Jupiter in the morning sky. The actual conjunction is at 06:08.*

Morning 7:00

10°

Mars

Spica 23

24

Saturn 25

26

Antares

S
SSW

January 23–26 • *The Moon passes Mars, Spica and Saturn.*

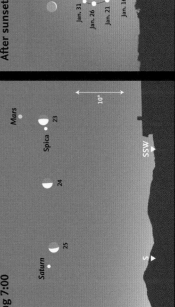

After sunset

10°

Feb. 01

Jan. 31
Jan. 26
Jan. 21
Jan. 16

Feb. 05
Jan. 31

Mercury

Sun

Evening apparition of Mercury • *Positions are shown in relation to the sun.*

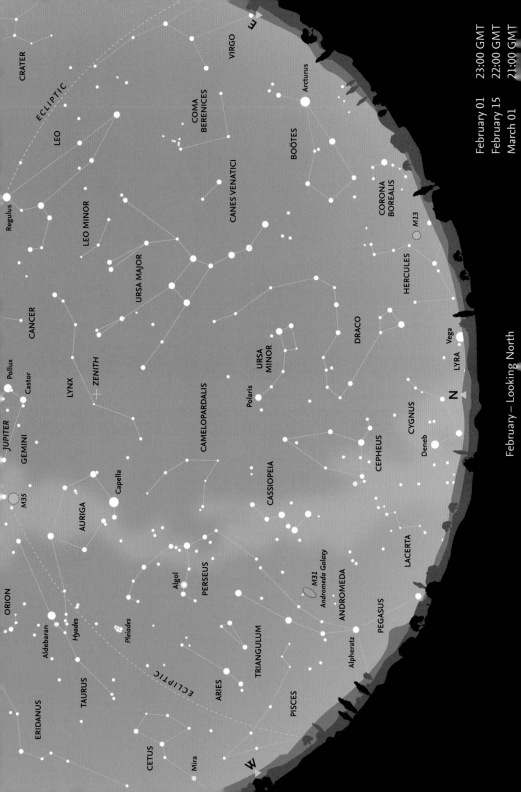

February 01   23:00 GMT
February 15   22:00 GMT
March 01   21:00 GMT

February – Looking North

February – Looking North

The months of January and February are probably the best time for seeing the section of the Milky Way that runs in the northern and western sky from **Cygnus**, low on the northern horizon, through **Cassiopeia**, **Perseus** and **Auriga** and then down through **Gemini** and **Orion**. Although not as readily perceived as the denser star clouds of the summer Milky Way, on a clear night so many stars may be visible that even a distinctive constellation such as **Cassiopeia** is not immediately obvious.

The head of **Draco** is now higher in the sky and easier to recognize. **Deneb**, (α Cygni), the brightest star in **Cygnus**, may just be visible almost due north at 24:00, early in the month, if the sky is very clear and the horizon clear of obstacles. **Vega** (α Lyrae) in **Lyra** is so low that it is difficult to see, but may become visible later in the night. The constellation of **Boötes** – sometimes described as shaped like a kite, an ice-cream cone, or the letter 'P' – with orange-tinted **Arcturus** (α Boötis), is beginning to clear the eastern horizon. Arcturus, at magnitude -0.05, is the brightest star in the northern hemisphere. The inconspicuous constellation of **Coma Berenices** is now well above the horizon in the east. The concentration of faint stars at the north-eastern corner somewhat resembles a tiny, detached portion of the Milky Way.

On the other side of the sky, in the north-west, most of the constellation of Andromeda is still easily seen, although **Alpheratz** (α Andromedae), the star that forms the north-eastern corner of the Great Square of **Pegasus** – even though it is actually part of Andromeda – is becoming close to the horizon and more difficult to detect.

High overhead, at the zenith, try to make out the very faint constellation of **Lynx**. It was introduced in 1687 by the famous astronomer Johannes Hevelius to fill the largely blank area between **Auriga**, **Gemini** and **Ursa Major**, and is reputed to be so named because one needed the eyes of a lynx to detect it.

In February the constellation of Boötes lies close to the north-eastern horizon. The brightest star in the photograph is orange-tinted Arcturus. The small constellation of Corona Borealis appears in the upper left-hand corner of the image.

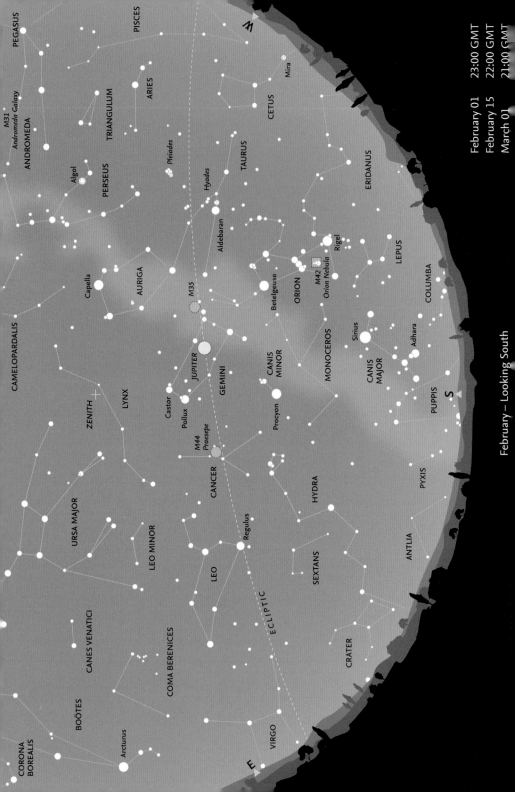

February – Looking South

February 01   23:00 GMT
February 15   22:00 GMT
March 01   21:00 GMT

February – Looking South

Apart from Orion, the most prominent constellation visible this month is **Gemini**, with its two lines of stars running south-east towards Orion. Many people have difficulty in remembering which–is–which of the two stars **Castor** and **Pollux**. Think of them in alphabetical order: Castor (α Gemini), the fainter star, is closer to the North Celestial Pole. Pollux (β Gemini) is the brighter of the two, but is farther away from the Pole. Castor is remarkable because it is actually a multiple system, consisting of no less than six individual stars.

Using Orion's belt as a guide, it points down to the south-east towards **Sirius**, the brightest star in the sky (at magnitude -1.4) in the constellation of **Canis Major**, the whole of which is now clear of the southern horizon. Forming a triangle with **Betelgeuse** in Orion and **Sirius** in Canis Major is **Procyon**, the brightest star in the small constellation of **Canis Minor**. Between Canis Major and Canis Minor is the faint constellation of **Monoceros**, which actually straddles the Milky Way, which is difficult to see in this area. Directly east of Procyon is the highly distinctive asterism of six stars that form the 'head' of **Hydra**, the largest of all 88 constellations, and which trails such a long way across

the sky that it is only in mid-March around midnight that the whole constellation becomes visible.

The constellation of Gemini: The two brightest stars are Castor and Pollux and can be found in the left part of the photograph.

The Moon's phases for February

01	02	03	04	05	06 19:22	07	08	09	10	11	12	13	14 23:53	15	16
					First Quarter										Full Moon

17	18	19	20	21	22 17:15	23	24	25	26	27	28
					Last Quarter						

February – Moon and Planets

The Moon

In the western sky on February 1, the Moon is close to *Mercury*. On February 7–8, it passes south of the *Pleiades* and then north of *Aldebaran* (both in Taurus). In the morning sky between February 19 and 22, it passes (in succession) *Spica*, *Mars* and *Saturn*, and is north of *Antares* on February 23, one day before Last Quarter (February 24). During this period, the Moon occults *Saturn* (on February 21), but this event is not visible from Britain. A further occultation (of *Venus*) occurs on February 26, and this time the occultation path is farther north (see the Calendar, next page). The diagram shows how (from Britain) Venus will appear north of the Moon on February 26. (In 2014, Saturn is actually occulted by the Moon in every month from January to October inclusive, but only the last of these is visible from any part of Europe.)

The Planets

Mercury sinks back towards the western horizon after greatest eastern elongation on January 31. *Venus* remains too close to the Sun to be easily (or safely) observed. *Mars* is moving eastwards in *Virgo* and becomes visible in the early morning. It begins the month at magnitude 0.3, but brightens considerably to end the month at magnitude -0.4. *Jupiter* is still retrograding in *Gemini* and is thus visible most of the night. At the beginning of the month it is at magnitude -2.6, but gradually fades to magnitude -2.4 by February 28 as its distance increases. *Saturn* is in *Libra*, visible in the morning sky, fading slightly from magnitude 0.5 to 0.4 over the course of the month.

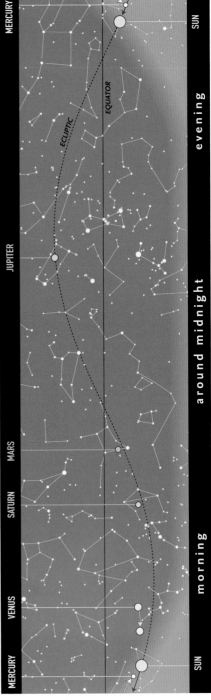

The path of the Sun and the planets along the ecliptic in February.

Calendar for February

Evening 17:45

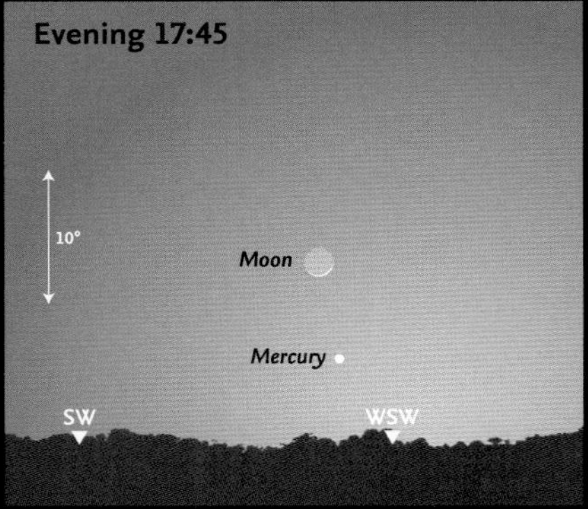

February 1 • The Moon and Mercury in the evening sky. The actual conjunction is at 07:07.

Evening 18:00

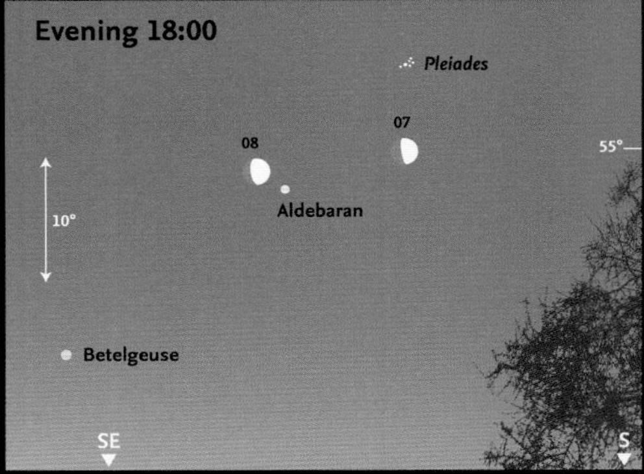

February 7–8 • The Moon passes the Pleiades and Aldebaran in the evening sky.

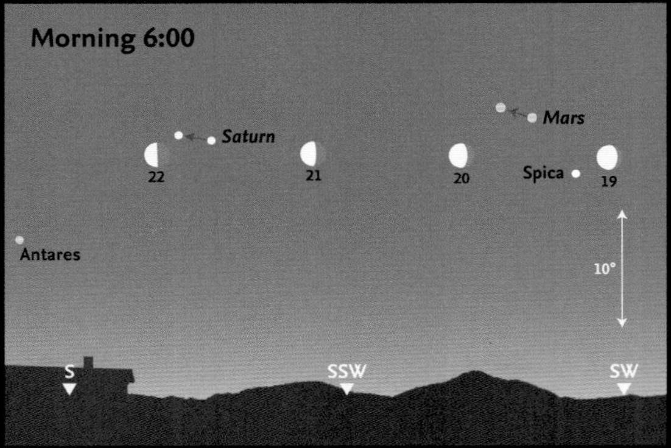

Morning 6:00

February 19–22 • The Moon passes Spica, Mars and Saturn.

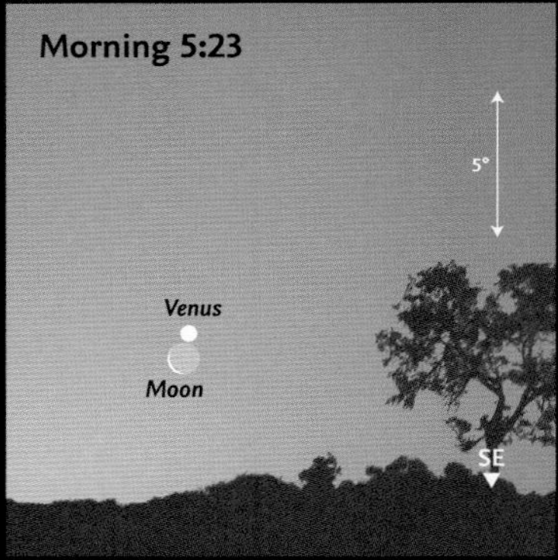

Morning 5:23

February 26 • Venus and the Moon close together in the morning sky.

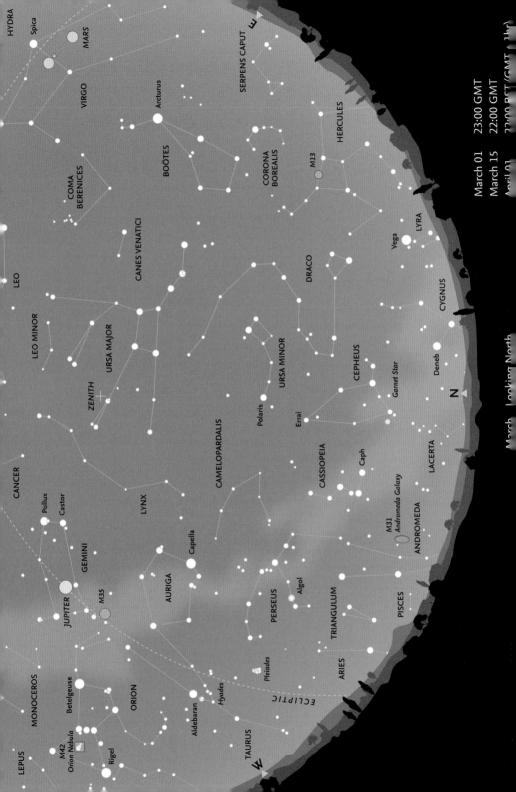

March 01   23:00 GMT
March 15   22:00 GMT
April 01   21:00 PST (GMT + 1hr)

March – Looking North

March – Looking North

In March, the Sun crosses the celestial equator on Thursday, March 20, at the vernal equinox, when day and night are of almost exactly equal length, and the season of spring is considered to have begun. (The hours of daylight and darkness change most rapidly around the equinoxes, in March and September.) It is also in March that Summer Time begins in Europe (on Sunday, March 30) so the charts show the appearance at 23:00 GMT for March 1 and 22:00 BST for April 1. (In the USA, Daylight Saving Time is introduced three weeks earlier, on Sunday, March 9.)

Early in the month, the constellation of *Cepheus* lies almost due north, with the distinctive 'W' of *Cassiopeia* to its west. Cepheus lies across the border of the Milky Way and is often described as like the gable-end of a house or a church tower and steeple. Despite the large number of stars revealed at the base of the constellation by binoculars, one star stands out because of its deep red colour. This is Mu (µ) Cephei, also known as the *Garnet Star*, because of its striking colour. It is a truly gigantic star, a red supergiant, and one of the largest stars known. It is about 2,400 times the diameter of the Sun, and if placed in the Solar System would extend beyond the orbit of Saturn. (Betelgeuse, in Orion, is also a red supergiant, but it is 'only' about 500 times the diameter of the Sun.)

Below Cepheus to the east, it may be possible to catch a glimpse of *Deneb* (α Cygni), just above the horizon. Slightly farther round towards the north-east, Vega (α Lyrae) is marginally higher in the sky. From southern Britain, Deneb is just far enough north to be circumpolar (although difficult to see in January and February because it is so low on the northern horizon). Vega, by contrast, farther south, is completely hidden during the depths of winter.

The brightest star on the right-hand side of this image is Polaris (α UMi). Cassiopeia, looking like a flattened letter 'M' is at top left. Between Polaris and Caph (β Cas) – the lowermost of the five stars – lies Errai (γ Cep) at the tip of the 'gable' or 'steeple'. At the base of Cepheus, among the stars of the Milky Way, the Garnet Star appears orange in this photograph.

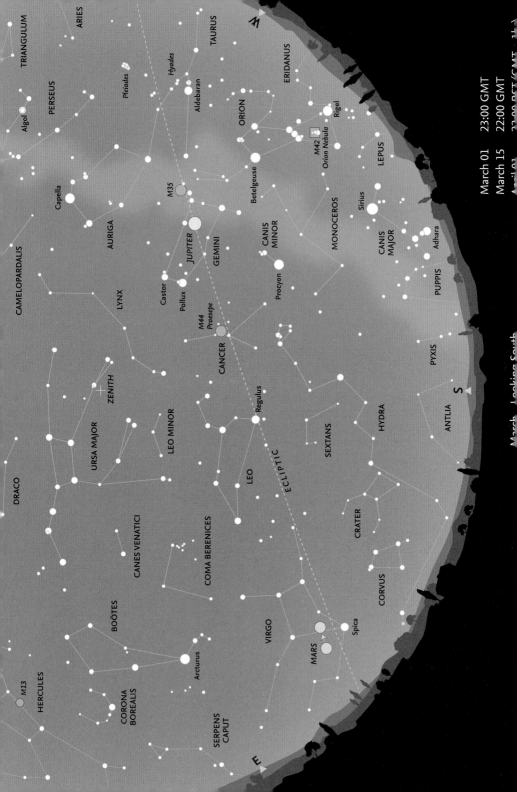

TRIANGULUM

ARIES

TAURUS

PERSEUS

Algol

Hyades

Aldebaran

Pleiades

ERIDANUS

CAMELOPARDALIS

Capella

ORION

Rigel

M42
Orion Nebula

AURIGA

Betelgeuse

LEPUS

M35

JUPITER

GEMINI

CANIS
MINOR

Sirius

MONOCEROS

LYNX

Castor

Pollux

Procyon

CANIS
MAJOR

Adhara

M44
Praesepe

PUPPIS

CANCER

DRACO

URSA MAJOR

ZENITH

LEO MINOR

Regulus

CANCER

PYXIS

HYDRA

SEXTANS

S

LEO

ANTLIA

ECLIPTIC

CANES VENATICI

COMA BERENICES

CRATER

BOÖTES

VIRGO

CORVUS

HERCULES

M13

Arcturus

CORONA
BOREALIS

MARS

Spica

SERPENS
CAPUT

E

W

March – Looking South

Due south at 22:00 at the beginning of the month, lying between the constellations of **Gemini** in the west and **Leo** in the east, and fairly high in the sky above the head of **Hydra**, is the faint, and rather undistinguished Zodiacal constellation of **Cancer**. Rather like the triskelion, the symbol for the Isle of Man, it has three 'legs' radiating from the centre, where there is an open cluster, M44 or *Praesepe* ('the Manger' but also known as 'the Beehive'). On a clear night this is just a hazy spot to the naked eye, but appears in binoculars as a group of dozens of individual stars.

Also prominent in March is the constellation of Leo, with the 'backward question-mark' (or 'Sickle') of bright stars forming the head of the mythological lion. **Regulus** (α Leonis) – the 'dot' of the 'question-mark' or the handle of the sickle and the brightest star in **Leo** – lies very close to the ecliptic and is one of the few first-magnitude stars that may be occulted by the Moon. The next occultation will not occur until 18 December 2016, visible only from southern Australia and part of Antarctica. A series of occultations follows in 2017, but not until 25 July 2017 and 8 December 2017 will occultations be visible from any part of Europe.

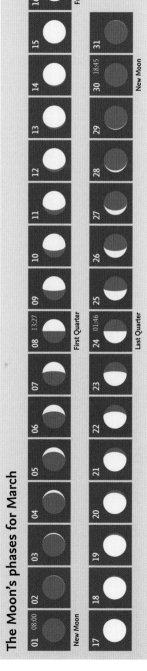

The constellation of Cancer lies between Castor and Pollux (on the right) and Regulus (on the left). To its south is the distinctive asterism of six stars marking the head of Hydra.

The Moon's phases for March

01 08:00	02	03	04	05	06	07	08 13:27
New Moon							First Quarter

09	10	11	12	13	14	15	16 17:08
							Full Moon

17	18	19	20	21	22	23	24 01:46
							Last Quarter

25	26	27	28	29	30 18:45	31
						New Moon

March – Moon and Planets

The Moon

On March 7, the six-day-old Moon passes close to *Aldebaran* (α Tauri). Later in the month, on March 18, it passes close to Mars and Spica (α Virginis) in the evening sky and, a few days later, on the morning of March 21 is located close to *Saturn*, which is actually occulted by the Moon for observers in the southern hemisphere. The next day (March 22) it is north of red *Antares* (α Scorpii).

The Planets

Although *Mercury* reaches greatest western elongation on March 14, this occurs after sunrise, and the planet is too close to the Sun to be visible this month. *Venus* comes to greatest western elongation of 47° on March 22. Although bright (magnitude -4.5) and fairly prominent, it is low on the eastern horizon before sunrise. *Mars* begins the month at mag. -0.5 and starts to retrograde (move westwards) the next day, March 2. It remains in *Virgo* throughout the month, rapidly brightening to mag. -1.3 by March 31. *Jupiter* remains readily visible in *Gemini* and ceases its retrograde motion on March 6, fading slowly from mag. -2.4 to -2.2 by the end of the month. *Saturn* spends the year in Libra, and is in the morning sky in March. It begins to retrograde on March 5, and increases slightly in brightness (from mag. 0.4 to 0.3) over the course of the month.

The path of the Sun and the planets along the ecliptic in March.

Calendar for March

01	08:00	New Moon
07	22:30	Aldebaran 2.1°S of Moon
08	13:27	First Quarter
09		Daylight Saving Time begins (USA)
10	11:27	Jupiter 5.2°N of Moon
11	17:10	Pollux 12.0°N of Moon
11	19:47	Moon at apogee
14	06:30	Mercury greatest elongation (28°W, mag. 0.1)
14	17:57	Regulus 5.1°N of Moon
16	17:08	Full Moon
18	21:00	Spica 1.7°S of Moon
19	03:14	Mars 3.2°N of Moon
20	16:57	Vernal equinox
21	03:18	Saturn 0.3°N of Moon
21		Saturn occulted by Moon (S. America, S. Atlantic, South Africa, Madagascar)
22	07:08	Antares 8.0°S of Moon
22	19:32	Venus greatest elongation (47°W, mag. -4.5)
24	01:46	Last Quarter
27	09:52	Venus 3.6°S of Moon
27	18:34	Moon at perigee
29	04:48	Mercury 6.2°S of Moon
30	18:45	New Moon
30		Summer Time begins (Europe)
31	04:00*	Spica 5.1°S of Mars

Entries with an * are at this distance for an extended period around this time.

Evening 22:30

March 7 • The Moon and Aldebaran close together in the evening sky.

Evening 22:00

March 18 • The Moon with Spica and Mars.

Morning 4:00

March 21–22 • The Moon passes Saturn and Antares.

Morning 4:00

March 31 • Spica with Mars and Saturn in the morning sky at 4:00 BST (Summer Time).

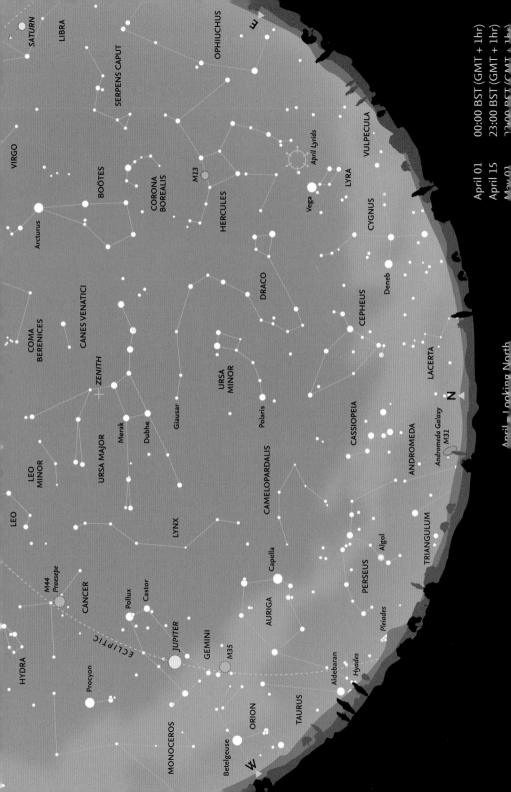

April – Looking North

April – Looking North

Cygnus and the brighter regions of the Milky Way are now becoming visible, rising in the north-east, with the small constellation of **Lyra** and the distinctive 'Keystone' of **Hercules** above them. The winding constellation of **Draco** weaves its way from the quadrilateral of stars that marks its 'head', on the border with Hercules, to end at **Giausar** (λ Draconis) between Polaris (α Ursae Minoris) and the 'Pointers', **Dubhe** and **Merak** (α and β Ursae Majoris, respectively). **Ursa Major** is high overhead, near the zenith. **Auriga** is still clearly seen in the north-west, but by the end of the month, the southern portion of **Perseus** is starting to dip below the northern horizon. The very faint constellation of **Camelopardalis** lies in the north-west between Polaris and the constellations of Auriga and Perseus.

Meteors

After the long period with no significant meteors showers, since the Quadrantids in January, a minor shower, the **Lyrids**, peaks on April 21. Although the hourly rate is not very high (about 15 meteors per hour), the meteors are fast and some leave persistent trails. This year, the maximum is four days before New Moon, so moonlight should not cause significant interference.

The constellation of Lyra is in the centre of this image, with the four stars forming the 'Keystone' of Hercules to the west (right) and the whole constellation of Cygnus visible to the east.

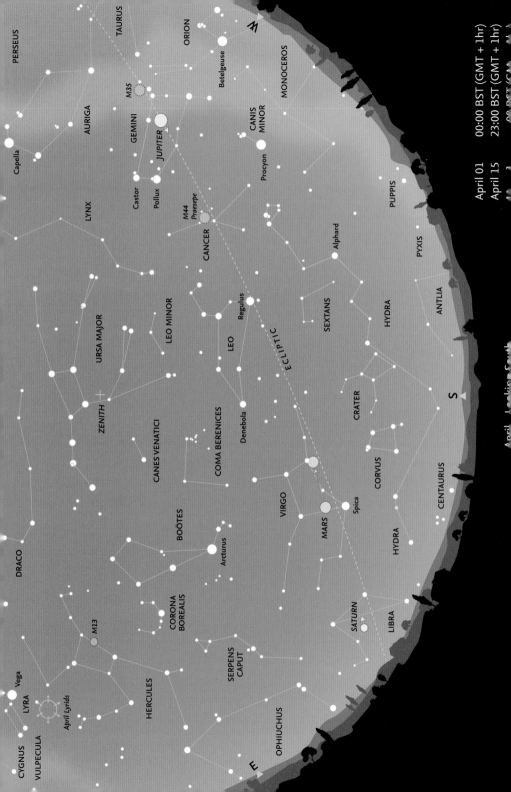

April 01   00:00 BST (GMT + 1hr)
April 15   23:00 BST (GMT + 1hr)

April – Looking South

Leo is the most prominent constellation in the southern sky in April, and vaguely looks like the creature after which it is named. **Gemini**, with **Castor** and **Pollux** remains clearly visible in the west, and **Cancer** lies between the two constellations. To the east of Leo, the whole of **Virgo**, with **Spica** (α Virginis) its brightest star, is well clear of the horizon. Below Leo and Virgo, the complete length of **Hydra** is visible, running below Leo and Virgo, with **Alphard** (α Hydrae) half-way between Regulus and the south-western horizon. Farther east, the two small constellations of **Crater** and the rather brighter **Corvus** lie between Hydra and Virgo.

Boötes and **Arcturus** are prominent in the eastern sky, together with the circlet of **Corona Borealis**, which is framed by Boötes and the neighbouring constellation of **Hercules**. Between Leo and Boötes lies the tiny constellation of **Coma Berenices**, most notable for being the location of the Coma Cluster of galaxies. There are about 1,000 galaxies in this cluster, which is located near the North Galactic Pole, where we are looking out of the plane of the Galaxy and are thus able to see deep into space. Only about ten of the brightest galaxies in the Coma Cluster are visible with the largest amateur telescopes.

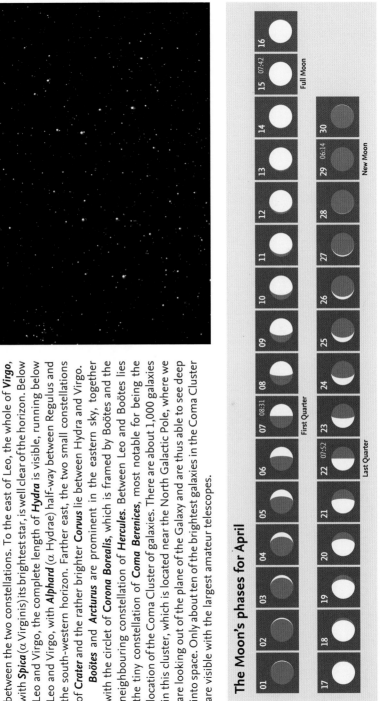

The highly distinctive constellation of Leo, with first-magnitude Regulus on the west, the 'Sickle' of stars above, and stretching east to the second-brightest star, Denebola.

The Moon's phases for April

01	02	03	04	05	06	07 08:31	08	09	10
11	12	13	14	15 07:42	16
17	18	19	20	21	22 07:52	23	24	25	26
27	28	29 06:14	30

First Quarter

Full Moon

Last Quarter

New Moon

April – Moon and Planets

The Moon

On April 6, the Moon passes across the triangle formed by *Jupiter* (in *Gemini*), *Betelgeuse* (in *Orion*) and *Procyon* (in *Canis Minor*). Full Moon occurs on April 15, when there is a total lunar eclipse, the whole of the Moon being obscured by the umbra (the darkest part of the Earth's shadow). The eclipse is visible from a wide area of the globe, although this is centred over the Pacific.

Over the period April 14–18, in the morning sky, the Moon passes *Mars*, *Spica*, and *Saturn*, and (on April 18) is in the same field of view as *Antares*. On April 25–26, the waning crescent Moon is close to *Venus* in the morning sky. A few days later, at New Moon on April 29, there is an annular eclipse of the Sun, when the Moon fails to cover the whole disk, and a narrow ring of the Sun remains visible around the Moon. The path of the eclipse lies mainly over the Southern Ocean, but is visible from eastern Australia and small portions of Antarctica. Mid-eclipse is at 06:03 UT (07:03 BST).

The Planets

Mercury is invisible, being above the horizon with the Sun throughout the month. *Venus* is low in the morning sky and might be seen under good conditions. It fades slighly from mag. -4.4 at the beginning of April to -4.2 at the end. *Mars* is still visible as it continues to retrograde in *Virgo*, and comes to opposition on April 8 at mag. -1.3. *Jupiter* is still in *Gemini*, with direct motion, and fades slightly from mag. -2.2 at the beginning of the month to mag. -2.0 at the end. *Saturn* becomes visible later in the night in *Libra* and it still retrograding. It brightens slightly from mag. 0.3 to mag. 0.1 over the course of the month.

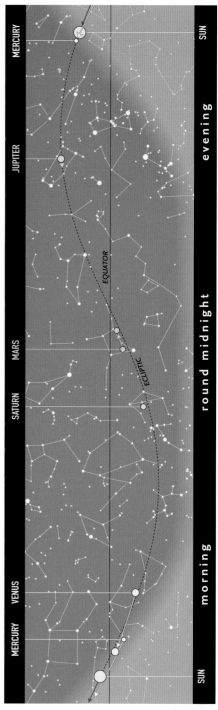

The path of the Sun and the planets along the ecliptic in April.

Calendar for April

04	07:14	Aldebaran 2.0°S of Moon
06	22:31	Jupiter 5.8°N of Moon
07	08:31	First Quarter
08	00:55	Pollux 12.2°N of Moon
08	16:42	Mars at oppostion (mag. −1.3)
08	14:52	Moon at apogee
11	01:50	Regulus 5.2°N of Moon
14	18:24	Mars 3.5°N of Moon
15	04:19	Spica 1.72°S of Moon
15	05:58	Total lunar eclipse (N. America, W. South America, Pacific, E. Australia)
15	07:42	Full Moon
17	07:20	Saturn 0.4°N of Moon
17		Saturn occulted by Moon (Pacific, S. South America, S.W. Atlantic)
18	12:56	Antares 8.0°S of Moon
19–25		April Lyrid meteor shower
21		April Lyrid meteor maximum
22	07:52	Last Quarter
23	00:24	Moon at perigee
Apr. 24 – May 20		Eta Aquarids
25	23:15	Venus 4.4°S of Moon
26	03:27	Mercury superior conjunction
29	06:03	Annular solar eclipse (E. Australia, Antarctica)
29	06:14	New Moon
29	13:04	Mercury 1.6°N of Moon

Evening 22:31

April 6 • *The Moon with Jupiter in the evening sky.*

Morning 5:30

April 25–26 • *The waning crescent Moon is close to Venus in the morning sky.*

Morning 4:00

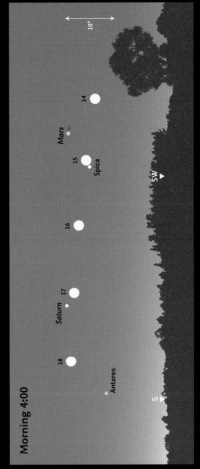

April 14–28 • *The Moon passes Mars, Spica, Saturn and Antares.*

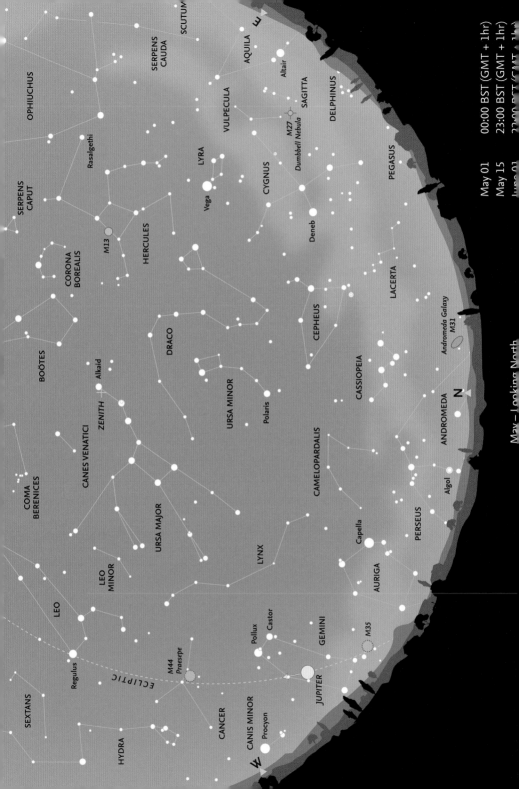

May – Looking North

May 01   00:00 BST (GMT + 1hr)
May 15   23:00 BST (GMT + 1hr)

May – Looking North

The constellation of Hercules. Rasalgethi (α Herculis) in the extreme south of the constellation, is near the bottom edge of this image.

Cassiopeia is now low over the northern horizon, and to its west, the southern portions of both **Perseus** and **Auriga** are becoming difficult to observe. **Gemini**, with **Castor** and **Pollux**, is sinking towards the western horizon.

In the east, two of the stars of the 'Summer Triangle', **Vega** in **Lyra**, and **Deneb** in **Cygnus** are clearly visible, and the third, **Altair** in **Aquila**, is beginning to climb above the horizon. The sprawling constellation of **Hercules** is high in the east and the brightest globular cluster in the northern hemisphere, M13, is visible to the naked eye on the eastern side of the 'Keystone'.

Later in the night (and in the month) the westernmost stars of **Pegasus** begin to come into view, while the stars of **Andromeda** are skimming the north-eastern horizon. High overhead, **Alkaid** (η Ursae Majoris), the last star in the 'tail' of the Great Bear is close to the zenith, while the main body of the constellation has swung round into the western sky.

Meteors

The **Eta Aquarids** are one of the two meteor showers associated with Comet P/Halley (the other being the **Orionids**, in October). The Eta Aquarids are not particularly favourably placed for northern-hemisphere observers, because the radiant is near the celestial equator, near the 'Water Jar' in Aquarius, well below the horizon until late in the night (around dawn). There is a radiant map for the Eta Aquarids on page 16.

Their maximum in 2014 occurs on May 5, a couple of days before First Quarter, so interference from moonlight is not particularly great. Quite a large proportion (about 25 per cent) of the meteors leave persistent trains.

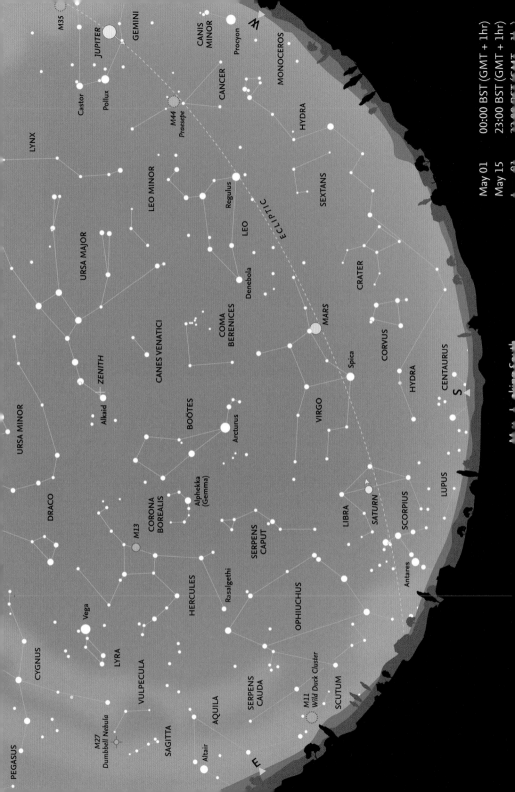

Looking South

May – Looking South

Early in the night, the constellation of *Virgo*, with *Spica* (α Virginis) lies due south, with *Leo* and both *Regulus* and *Denebola* (α and β Leonis, respectively) to its east still well clear of the horizon. Later in the night, the rather faint zodiacal constellation of *Libra* becomes visible, and to its east, the ruddy star *Antares* (α Scorpii) begins to climb up over the horizon.

Arcturus in *Boötes* is high in the south, with the distinctive circlet of *Corona Borealis* clearly visible to its east. The brightest star (α Coronae Borealis) is known as *Alphekka* or *Gemma*. The large constellation of *Ophiuchus* (which actually crosses the ecliptic, and is thus the 'thirteenth' zodiacal constellation) is climbing into the eastern sky. Before the constellation boundaries were formally adopted by the International Astronomical Union in 1930, the southern region of Ophiuchus was regarded as forming part of the constellation of Scorpius, which had been part of the Zodiac since antiquity.

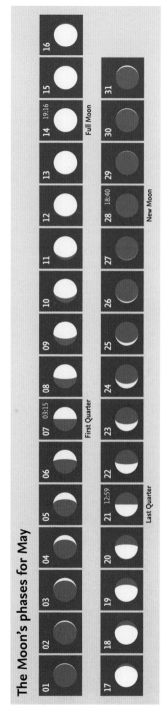

The constellation of Virgo. The brightest object is the planet Mars, which is again in this constellation for the first half of 2014.

The Moon's phases for May

| 01 | 02 | 03 | 04 | 05 12:59 | 06 | 07 03:15 | 08 | 09 | 10 | 11 | 12 | 13 | 14 19:16 | 15 | 16 |

Last Quarter — First Quarter — Full Moon

| 17 | 18 | 19 | 20 | 21 12:59 | 22 | 23 | 24 | 25 | 26 | 27 | 28 18:40 | 29 | 30 | 31 |

New Moon

May – Moon and Planets

The Moon

On May 1, the waxing crescent Moon passes close to Aldebaran in the western sunset sky. In the days just before Full Moon (on May 14), the Moon again passes, in sequence, *Mars* and *Spica* (in Virgo) and *Saturn* (in Libra). Later in the month on successive days (May 25–26) it is close to *Venus* in the early-morning sky.

The Planets

Mercury may be glimpsed in the western sky after sunset as it comes to greatest eastern elongation on May 25 at mag. 0.3, but it fades rapidly to mag. 1.1 by May 31. *Venus* is very low in the early-morning sky and difficult to see, despite it starting the month at mag. -4.1 and only fading very slightly (to mag. -4.0) by May 31. *Mars* remains

in Virgo, and is thus visible for most of the night. It resumes direct motion on May 21 (after opposition on April 6), but fades fairly rapidly from mag. -1.2 to mag. -0.5 over the course of the month. *Jupiter*, is still in Gemini, with direct motion, and fades very slightly over the month from mag. -2.0 to -1.9. *Saturn* is visible throughout the night as it continues to show retrograde motion in *Libra*, slowly fading from mag. 0.1 to mag. 0.2 over the course of the month.

The path of the Sun and the planets along the ecliptic in May.

1	16:13	Aldebaran 2.0°S of Moon
5		Eta Aquarid maximum
05	09:01	Pollux 12.1°N of Moon
06	10:23	Moon at apogee
07	03:15	First Quarter
08	10:07	Regulus 5.1°N of Moon
10	18:28	Saturn at opposition (mag. +0.3)
12	13:09	Spica 1.7°S of Moon
13	16:00*	Aldebaran 7.8°S of Mercury
14	19:16	Full Moon
14		Saturn occulted by Moon (Australia, New Zealand, Antarctica, S. Pacific)
15	20:36	Antares 8.1°S of Moon
18	11:57	Moon at perigee
21	12:59	Last Quarter
25	07:10	Mercury greatest elongation (23°E, mag. 0.4)
28	18:40	New Moon
29	00:13	Aldebaran 2.0°S of Moon

Entries with an * are at this distance for an extended period around this time.

Evening 21:00

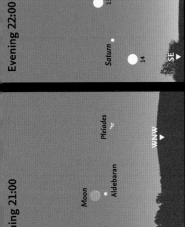

Moon
Aldebaran
Pleiades

10°

W   WNW

May 1 • The waxing crescent Moon close to Aldebaran in the evening sky.

Evening 22:00

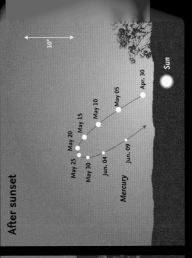

10°

Mars   10
11
Spica
12
13
Saturn
14

S   SSE   SE

May 10–14 • The Moon passes Mars, Spica and Saturn.

After sunset

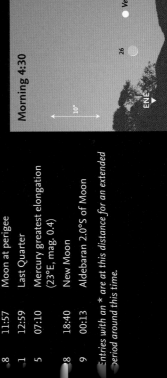

10°

May 20
May 25
May 15
May 10
May 30
May 05
Apr. 30
Jun. 04
Jun. 09
Mercury
Sun

Evening apparition of Mercury. Positions are shown ...

Morning 4:30

10°

25
26
Venus
E
ENE

May 25–26 • The Moon and Venus in the morning sky.

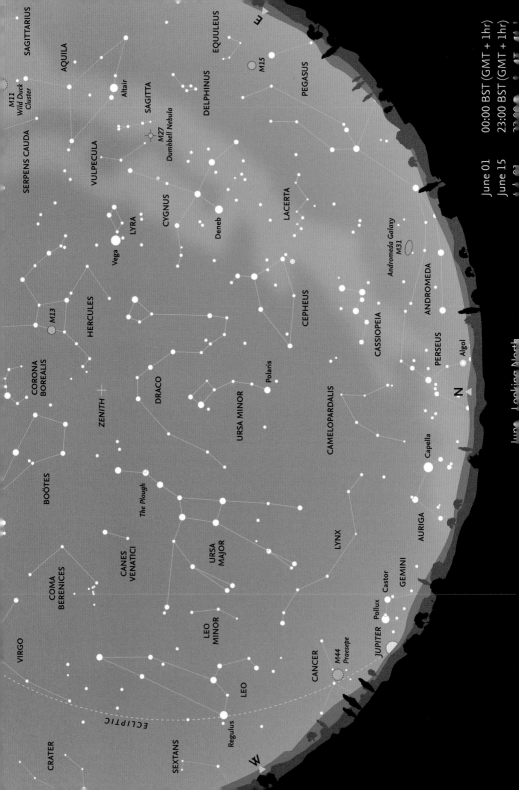

June – Looking North

June 01	00:00 BST (GMT + 1hr)
June 15	23:00 BST (GMT + 1hr)

June – Looking North

As we approach the summer solstice (June 21), even in southern England and Ireland a form of twilight persists throughout the night, and farther north, in Scotland, the sky remains so light that most of the fainter stars and constellations are invisible. Even brighter stars, such as the seven stars making up the well-known asterism known as the *Plough* in **Ursa Major** may be difficult to detect except around local midnight, 00:00 UT (01:00 BST).

But there is one compensation during these light nights: even southern observers may be lucky enough to witness a display of noctilucent clouds (NLC). These are highly distinctive clouds shining with an electric-blue tint, observed in the sky in the direction of the North Pole. They are the highest clouds in the atmosphere, occurring at altitudes of 80–85 km, far above all other clouds. They are only visible during summer nights, for about a month or six weeks on either side of the solstice, when observers are in darkness, but the clouds themselves remain illuminated by sunlight, reaching them from the Sun, itself hidden below the northern horizon.

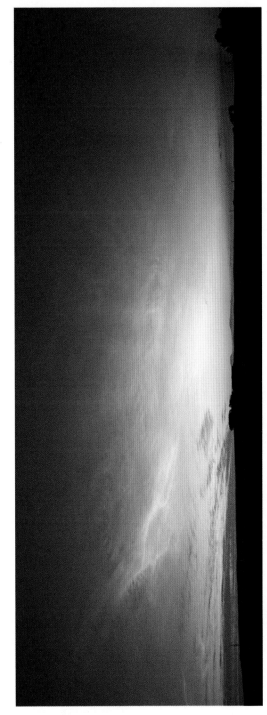

Noctilucent clouds consist of ice particles that have formed on tiny specks of meteoric dust that have entered the atmosphere from interplanetary space.

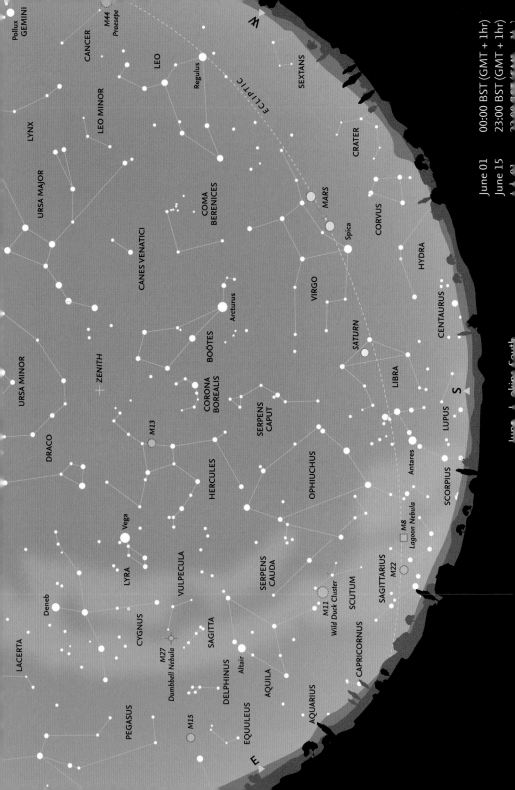

June – Looking South

Pollux
GEMINI
M44
Praesepe
CANCER
LEO
LEO MINOR
LYNX
Regulus
SEXTANS
CRATER
ECLIPTIC
URSA MAJOR
COMA
BERENICES
CANES VENATICI
MARS
Spica
CORVUS
HYDRA
VIRGO
ZENITH
URSA MINOR
DRACO
Arcturus
BOÖTES
CORONA
BOREALIS
SERPENS
CAPUT
SATURN
CENTAURUS
M13
LIBRA
S
HERCULES
OPHIUCHUS
LUPUS
Antares
SCORPIUS
Vega
LYRA
VULPECULA
SERPENS
CAUDA
M8
Lagoon Nebula
Deneb
CYGNUS
SAGITTA
SCUTUM
M22
SAGITTARIUS
LACERTA
M27
Dumbbell Nebula
DELPHINUS
Altair
AQUILA
M11
Wild Duck Cluster
CAPRICORNUS
PEGASUS
EQUULEUS
M15
AQUARIUS
E

June – Looking South

Although the persistent twilight makes observing even the southern sky difficult, the rather undistinguished constellation of **Libra** lies almost due south. The red supergiant star **Antares** – the name means the 'Rival of Mars' – in **Scorpius** is visible slightly to the east of the meridian, but the 'tail' or 'sting' remains below the horizon. Higher in the sky is the large constellation of **Ophiuchus** (the 'Serpent Bearer'), lying between the two halves of the constellation of **Serpens: Serpens Caput** ('head of the serpent') to the west, and **Serpens Cauda** ('tail of the serpent') to the east. (Serpens is the only constellation to be divided into two distinct parts.) The ecliptic runs across Ophiuchus, and the Sun actually spends far more time in the constellation than it does in the 'classical' zodiacal constellation of Scorpius, a small area of which lies between Libra and Ophiuchus.

Higher in the southern sky, the three constellations of **Boötes**, **Corona Borealis** and **Hercules** are now better-placed for observation than at any other time of the year.

The constellation of Ophiuchus.

The Moon's phases for June

01	02	03	04	05 20:39	06	07	08
				First Quarter			
09	10	11	12	13 04:11	14	15	16
						Full Moon	
17	18	19 18:39	20	21	22	23	24
		Last Quarter					
25	26	27 08:08	28	29	30		
		New Moon					

June – Moon and Planets

The Moon

Early in the month (June 4), the Moon passes south of Regulus in the western sky. Then, in the days between June 7 and 12, leading up to Full Moon (June 13), the southern part of the sky shows the Moon again passing a series of planets and stars: *Mars*, *Spica*, and *Saturn*, with the addition, this month, of *Antares* (on the night before Full Moon). On June 24 and 25, the Moon is close to the *Pleiades*, *Venus*, and *Aldebaran* in the early-morning sky.

The Planets

Mercury passes inferior conjunction (between Earth and Sun) on June 19, so remains invisible this month. *Venus*, at mag. -4.0 to -3.9 during the month, is very low in the morning sky, but may be visible towards the end of the month. *Mars*, having resumed direct motion in late May, is now moving slowly towards the east, but is still within the constellation of *Virgo*. It fades from mag. -0.5 at the beginning of the month to mag. 0.0 at the end. *Jupiter* is moving eastwards out of *Gemini* towards *Cancer*, but is very low in the western sky after sunset and difficult to detect. It fades very slightly over the month, beginning at magnitude -1.9 and decreasing to just -1.8 by June 30. *Saturn* remains visible in *Libra*, the constellation in which it lies for nearly the whole year, leaving it to enter Scorpius only in late December. Over the month of June it fades very slightly from magnitude 0.2 to 0.3.

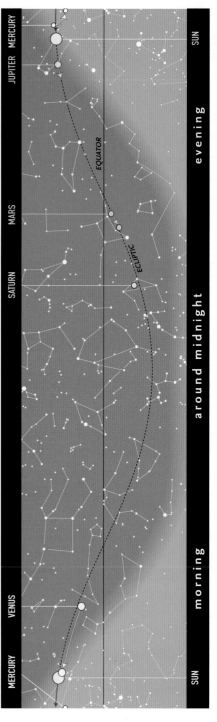

The path of the Sun and the planets along the ecliptic in June.

Calendar for June

01	16:48	Pollux 12.0°N of Moon
03	04:25	Moon at apogee
04	17:58	Regulus 5.0°N of Moon
05	20:39	First Quarter
08	22:27	Spica 1.8°S of Moon
10		Saturn occulted by Moon (S. Atlantic, S. Africa, S. Indian Ocean, Antarctica, extreme SW Australia)
12	06:07	Antares 8.0°S of Moon
13	04:11	Full Moon
15	03:29	Moon at perigee
19	21:50	Mercury inferior conjunction
19	18:39	Last Quarter
21	10:51	Summer solstice
21	12:00*	Pollux 6.4°N of Jupiter
25	06:45	Aldebaran 2.0°S of Moon
26		Mercury occulted by Moon (SE North America, Central America, NW South America, Atlantic, N. Africa, Central & S. Europe, Arabia)
27	08:08	New Moon
28	23:45	Pollux 11.9°N of Moon
30	19:10	Moon at apogee

Entries with an ★ are at this distance for an extended period around this time.

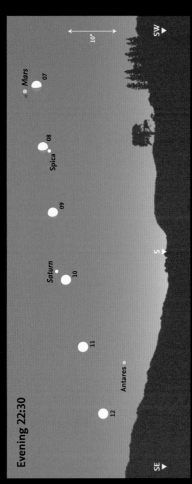

Evening 22:30

June 7–12 • *The Moon passes Mars, Spica, Saturn and Antares.*

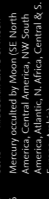

Evening 22:30

June 4 • *The Moon and Regulus high in the evening sky. The actual conjunction occurs in daylight (18:58 BST).*

Morning 4:30

June 24–25 • *The Moon passes the Pleaides, Venus and Aldebaran.*

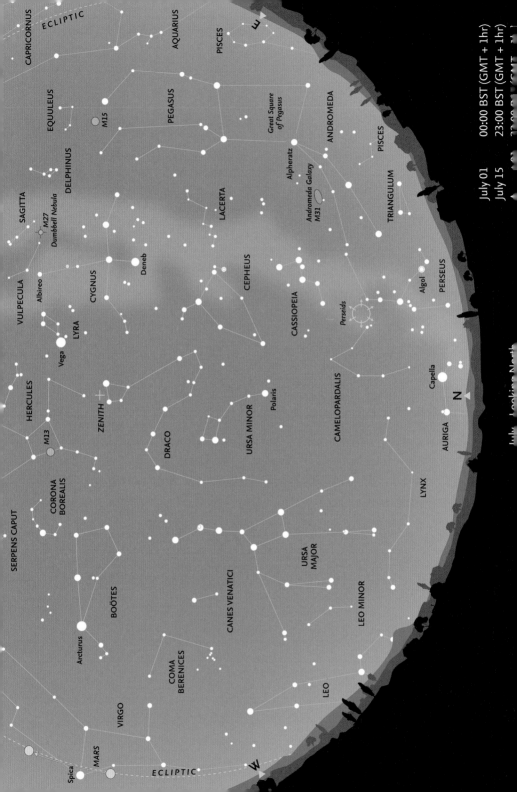

CAPRICORNUS

ECLIPTIC

AQUARIUS

PISCES

EQUULEUS

M15

PEGASUS

DELPHINUS

Great Square
of Pegasus

SAGITTA

M27
Dumbbell Nebula

ANDROMEDA

LACERTA

Alpheratz

PISCES

Andromeda Galaxy
M31

VULPECULA

Albireo

CYGNUS

Deneb

CEPHEUS

TRIANGULUM

LYRA

Vega

CASSIOPEIA

Algol

PERSEUS

Perseids

HERCULES

ZENITH

DRACO

URSA MINOR

Polaris

CAMELOPARDALIS

Capella

M13

CORONA
BOREALIS

N

SERPENS CAPUT

AURIGA

LYNX

BOÖTES

Arcturus

CANES VENATICI

URSA
MAJOR

COMA
BERENICES

LEO MINOR

VIRGO

LEO

MARS

Spica

ECLIPTIC

W

July, Looking North

July 01   00:00 BST (GMT + 1hr)
July 15   23:00 BST (GMT + 1hr)

July – Looking North

As in June, light nights and the chance of observing noctilucent clouds persist throughout July, but later in the month (and particularly after midnight) some of the major constellations begin to be more easily seen. **Capella**, the brightest star in **Auriga** (most of which is too low to be visible) is skimming the northern horizon. **Cassiopeia** is clearly visible in the north-east and **Perseus**, to its south, is beginning to climb clear of the horizon. The band of the Milky Way from Perseus, through Cassiopeia towards **Cygnus** stretches up into the north-eastern sky. If the sky is dark and clear, you may be able to make out the small, faint constellation of **Lacerta**, lying across the Milky Way between Cassiopeia and Cygnus. In the east, the stars of **Pegasus** are now well clear of the horizon, with the main line of stars forming **Andromeda** roughly parallel to the horizon in the north-east. **Alpheratz** (α Andromedae) is actually the star at the north-eastern corner of the Great Square of **Pegasus**. **Cepheus** and **Ursa Major** are on opposite sides of **Polaris** and **Ursa Minor,** in the east and west, respectively. The head of **Draco** is very close to the zenith so the whole of this winding constellation is readily seen.

Meteors

July brings increasing meteor activity, mainly because there are several minor radiants active in the constellations of **Capricornus** and **Aquarius**. Because of their location, however, observing conditions are not favourable for northern-hemisphere observers, and the most prominent shower is probably that of the **Delta Aquarids**, which are active from around July 15 to August 20, with a peak on July 29, although even then the hourly rate is unlikely to reach 25 meteors per hour. The radiant is shown on the star map for August, looking south. This year, maximum is just three days after New Moon, so observational conditions are favourable. Unfortunately, the same cannot be said for the **Perseids**, where shower activity begins on July 15, but does not peak until August 20.

Cygnus, sometimes known as the 'Northern Cross' depicts a swan, flying down the Milky Way, towards Sagittarius. The brightest star, Deneb (α Cygni) represents the tail, and Albireo (β Cygni) marks the position of the head, and may be found in the lower right section of the image.

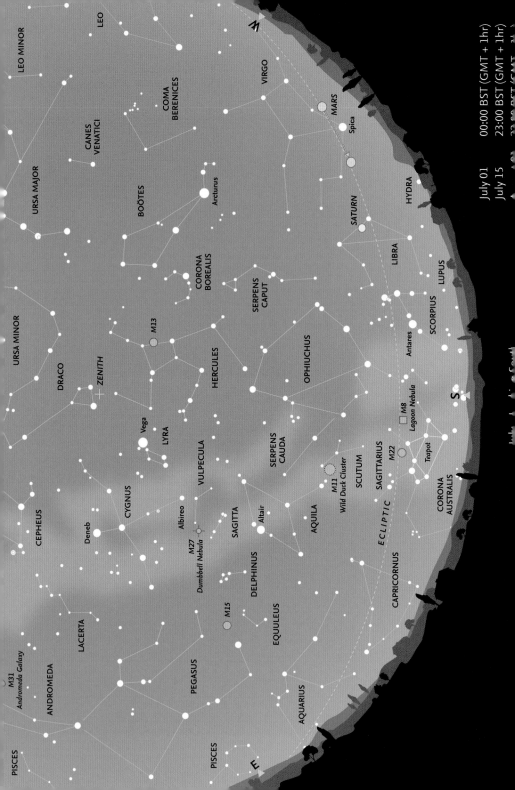

July Looking South

July 01   00:00 BST (GMT + 1hr)
July 15   23:00 BST (GMT + 1hr)

July – Looking South

Although part of the constellation remains hidden, this is perhaps the best time of year to see **Scorpius**, with deep red **Antares** (α Scorpii), glowing just above the southern horizon. At around midnight (UT), 01:00 BST, part of **Sagittarius** with the distinctive asterism of the 'Teapot', and the dense star clouds of the centre of the Milky Way are just visible in the south. With clear skies, the Great Rift – actually dust clouds that hide the more distant stars – runs down the Milky Way from Cygnus towards Sagittarius. The sprawling constellation of **Ophiuchus** lies close to the meridian for a large part of the month, separating the two halves of the constellation of **Serpens.** The western half is called **Serpens Caput** (Head of the Serpent) and the eastern part **Serpens Cauda** (Tail of the Serpent). In the east, the bright **Summer Triangle**, consisting of **Vega** in **Lyra**, **Deneb** in **Cygnus** and Altair in **Aquila** begins to dominate the southern sky, as it will throughout August and into September. The small constellation of Lyra, with Vega and a distinctive quadrilateral of stars to its east and south, lies not far south of the zenith.

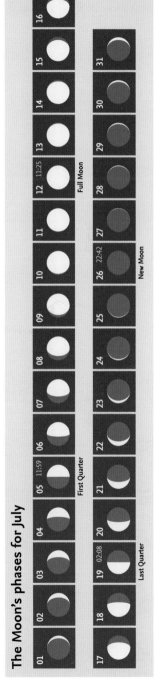

The Summer Triangle, consiting of the brightest stars in three constellations (Cygnus, Lyra and Aquila), is a constant feature of the night skies in summer, just as the constellation of Orion is during the winter.

The Moon's phases for July

July – Moon and Planets

The Moon

Once again, as in previous months, over the nights July 5–9, the Moon passes a succession of objects in the sky: first *Mars* and *Spica*, then *Saturn* and finally, *Antares*. (For observers in the southern hemisphere, the Moon occults both Mars and Saturn on July 6 and July 8, respectively.) Later in the month, just before New Moon (July 26) between July 22 and 24, the Moon is seen close to *Mercury* and *Venus* in the morning sky. Two days after New Moon, on July 28, the Moon passes apogee (the farthest point in its orbit around the Earth). Its actual distance then at 406,566.842 km is the greatest to occur in 2014.

The Planets

Mercury comes to greatest western elongation on July 12 at mag. 0.3. *Venus* remains bright (mag. -3.9 to mag. -3.8) and is visible around dawn in the early-morning twilight. It is close to Mercury in the dawn sky for most of the month. *Mars* continues its direct motion and passes close to *Spica* on July 12. It fades from magnitude 0.0 at the beginning of the month to 0.4 at the end. *Jupiter* is too close to the Sun to be visible in July. It is in conjunction with the Sun on July 24. Initially, *Saturn* is still an evening object, retrograding in Libra, but it reaches a stationary point on July 22, and resumes direct motion after that date. It fades very slightly during the month from mag. 0.4 to 0.5.

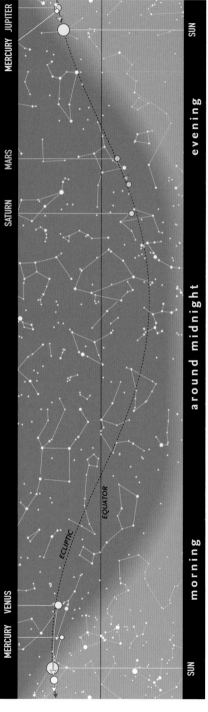

The path of the Sun and the planets along the ecliptic in July.

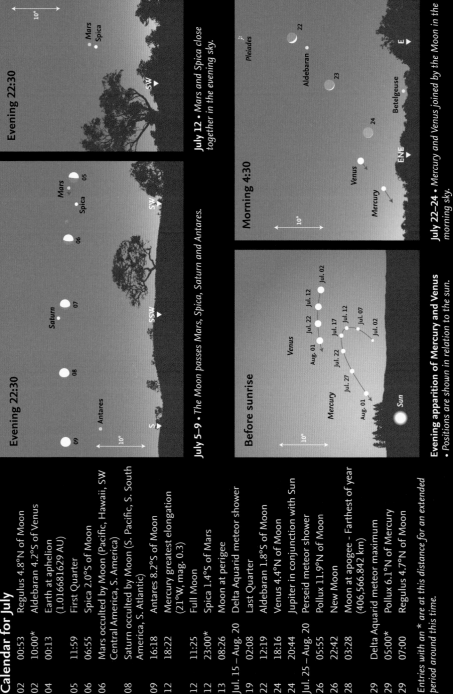

Calendar for July

02	00:53	Regulus 4.8°N of Moon
02	10:00*	Aldebaran 4.2°S of Venus
04	00:13	Earth at aphelion (1.01668629 AU)
05	11:59	First Quarter
06	06:55	Spica 2.0°S of Moon
06		Mars occulted by Moon (Pacific, Hawaii, SW Central America, S. America)
08		Saturn occulted by Moon (S. Pacific, S. South America, S. Atlantic)
09	16:18	Antares 8.2°S of Moon
12	18:22	Mercury greatest elongation (21°W, mag. 0.3)
12	11:25	Full Moon
12	23:00*	Spica 1.4°S of Mars
13	08:26	Moon at perigee
Jul. 15 – Aug. 20		Delta Aquarid meteor shower
19	02:08	Last Quarter
22	12:19	Aldebaran 1.8°S of Moon
24	18:16	Venus 4.4°N of Moon
24	20:44	Jupiter in conjunction with Sun
Jul. 25 – Aug. 20		Perseid meteor shower
26	05:55	Pollux 11.9°N of Moon
26	22:42	New Moon
28	03:28	Moon at apogee - Farthest of year (406,566.842 km)
29		Delta Aquarid meteor maximum
29	05:00*	Pollux 6.1°N of Mercury
29	07:00	Regulus 4.7°N of Moon

Entries with an ★ are at this distance for an extended period around this time.

Evening 22:30

July 5–9 • *The Moon passes Mars, Spica, Saturn and Antares.*

Evening 22:30

July 12 • *Mars and Spica close together in the evening sky.*

Before sunrise

Evening apparition of Mercury and Venus
• *Positions are shown in relation to the sun.*

Morning 4:30

July 22–24 • *Mercury and Venus joined by the Moon in the morning sky.*

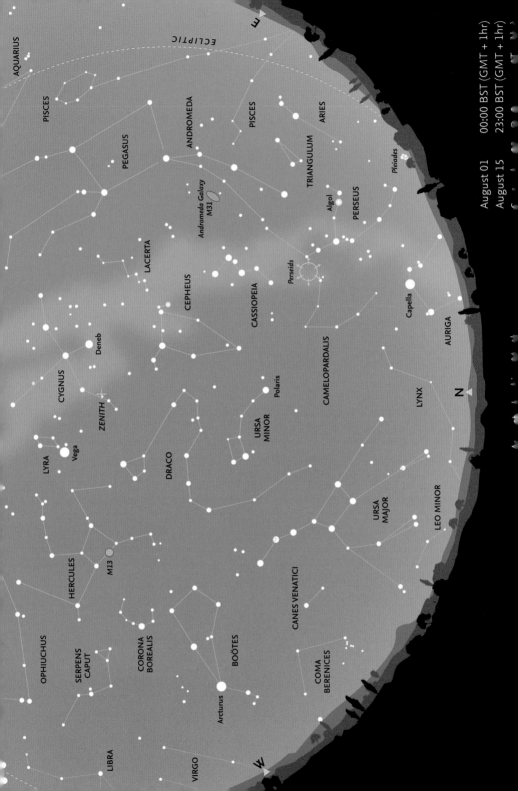

AQUARIUS

ECLIPTIC

PISCES

PEGASUS

ANDROMEDA

PISCES

TRIANGULUM

ARIES

LACERTA

Andromeda Galaxy
M31

CEPHEUS

CASSIOPEIA

Algol

Pleiades

PERSEUS

Perseids

Capella

Deneb

AURIGA

CYGNUS

CAMELOPARDALIS

ZENITH

Polaris

N

LYRA

Vega

URSA
MINOR

DRACO

LYNX

HERCULES

URSA
MAJOR

M13

LEO MINOR

OPHIUCHUS

CANES VENATICI

SERPENS
CAPUT

CORONA
BOREALIS

BOÖTES

COMA
BERENICES

LIBRA

Arcturus

VIRGO

W

E

August 01   00:00 BST (GMT + 1hr)
August 15   23:00 BST (GMT + 1hr)

August – Looking North

Ursa Major is now the 'right way up' in the north-west, although some of the fainter stars in the south of the constellation are difficult to see. Beyond it, **Boötes** stands almost vertically in the west, but pale orange **Arcturus** is sinking towards the horizon. Higher in the sky, both **Corona Borealis** and **Hercules** are clearly visible.

In the north-east, **Capella** is clearly visible, but most of **Auriga** still remains below the horizon. Higher in the sky, **Perseus** is gradually coming into full view, and later in the night and later in the month, the beautiful **Pleiades** open cluster rises above the north-eastern horizon. Between Perseus and **Polaris** lies the faint, and unremarkable constellation of **Camelopardalis**.

Higher still, both **Cassiopeia** and **Cepheus** are well placed for observation, despite the fact that Cassiopeia is completely immersed in the band of the Milky Way, as is the 'base' of Cepheus. **Pegasus** and **Andromeda** are now well above the eastern horizon and, below them, the constellation of **Pisces** is climbing into view. Two of the stars in the **Summer Triangle**, **Deneb** and **Vega**, are close to the zenith high overhead.

Meteors

August is regarded by astronomers as the month when one of the best meteor showers of the year occurs: the **Perseids**. This is a long shower, generally beginning about July 20 and continuing until around August 20, with a maximum on August 12, when the rate may reach as high as 80 meteors per hour (and on rare occasions, even higher). Unfortunately, in 2014, maximum is just 2 days after Full Moon, so interference by moonlight will make observing difficult. The Perseids are debris from Comet P/Swift-Tuttle (the Great Comet of 1862). Perseid meteors are fast and many of the brighter ones leave persistent trains.

The constellation of Perseus, is not only the location of the radiant for the Perseid meteor shower, but is also well-known for the pair of open clusters, known as the Double Cluster (near the top edge of the image) close to the border with Cassiopeia, and also for Algol (β Persei), a famous variable star that may be found at the centre of the photograph. The open star cluster, the Pleiades, in Taurus, is close to the bottom edge of the image.

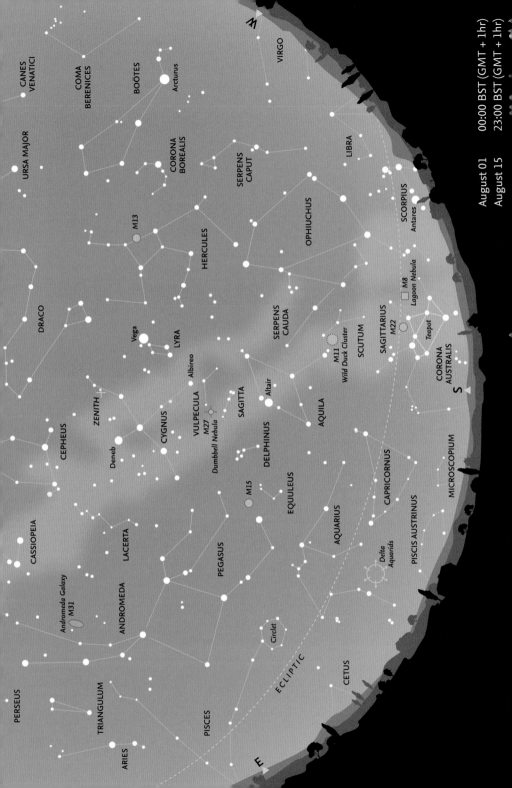

August 01   00:00 BST (GMT + 1hr)
August 15   23:00 BST (GMT + 1hr)

August – Looking South

The whole stretch of the summer Milky Way stretches across the sky in the south, from **Cygnus**, high in the sky near the zenith, past **Aquila**, with bright **Altair** (α Aquilae) to part of the constellation of **Sagittarius** close to the horizon, where the pattern of stars known as the 'Teapot' may be just visible. Between **Albireo** (β Cygni) and Altair lie the two small constellations of **Vulpecula** and **Sagitta**, with the latter easier to distinguish (because of its shape) from the clouds of the Milky Way. Between Sagitta and Pegasus to the east, lie the highly distinctive five stars that form the tiny constellation of **Delphinus** (again, one of the few constellations that actually bear some resemblance to the creatures after which they are named). Below Aquila, mainly in the star clouds of the Milky Way, lies **Scutum**, most famous for the bright open cluster, M11 or the 'Wild Duck Cluster', readily visible in binoculars. To the south-east of Aquila lie the two zodiacal constellations of **Capricornus** and **Aquarius**.

Aquila, with Altair (α Aquilae), its brightest star, like Cygnus, represents a bird – in this case, an eagle – flying down the length of the Milky Way.

The Moon's phases for August

01	02	03	04 00:50	05	06	07	08	09	10 18:09	11	12	13	14	15	16

First Quarter (04) Full Moon (10)

17 02:26	18	19	20	21	22	23	24	25 14:13	26	27	28	29	30	31

Last Quarter (17) New Moon (25)

August – Moon and Planets

The Moon

On August 10, the Moon is at perigee (closest to the Earth in its orbit) as it is every month. This time it is the closest approach of the year, when it is 356,896 km distant, which may be compared with its farthest distance (on July 28). Although reporters often make a great fuss about such occasions in the media, the difference in size is very small, and not perceptible to the human eye. On August 23 and 24, the waning crescent Moon joins Venus and Jupiter in the early-morning sky just before dawn. On the very last day of the month, it passes just south of Saturn in the triangle formed by Saturn, Mars, and Spica. (There is an occultation of Saturn for observers in the southern United States, northern South America and South Africa.)

The Planets

Mercury is once again at superior conjunction (on the far side of the Sun) on August 8, so it is not visible at any time during the month. *Venus* is low in the eastern pre-dawn sky. It is at magnitude -3.8 for most of August, but brightens slightly to mag. -3.9 at the end of the month. Over the course of August, *Mars* finally leaves *Virgo* and enters *Libra*. Beginning the month at mag. 0.4 it fades slowly to mag. 0.6 at the end of the month. *Jupiter* is in *Cancer*, but remains invisible until the very end of August, when it may perhaps be glimpsed in the early dawn sky. *Saturn* is moving slowly eastwards in Libra and appears close to Mars in the latter half of the month, when both planets are at magnitude 0.6.

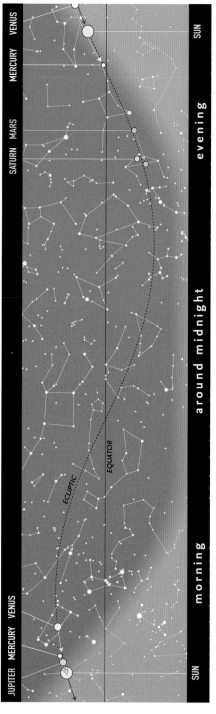

The path of the Sun and the planets along the ecliptic in August.

Calendar for August

02	13:50	Spica 2.3°S of Moon
02	17:00*	Jupiter 1.0°S of Mercury
04	00:50	First Quarter
04		Saturn occulted by Moon (S. India, Indian Ocean, S. Indonesia, Australia)
06	01:36	Antares 8.4°S of Moon
07	21:00*	Pollux 6.6°N of Venus
08	16:21	Mercury superior conjunction
10	17:43	Moon at perigee (Closest of year = 356,895,595 km)
10	18:09	Full Moon
12		Perseid meteor maximum
15	11:00*	Regulus 1.3°S of Mercury
17	12:26	Last Quarter
18	04:00*	Jupiter 0.2°S of Venus
18	18:09	Aldebaran 1.6°N of Moon
22	11:50	Pollux 12.0°N of Moon
24	06:09	Moon at apogee
25	12:53	Regulus 4.6°N of Moon
25	14:13	New Moon
27	13:00*	Saturn 3.6°N of Mars
29	19:31	Spica 2.5°S of Moon
31		Saturn occulted by Moon (SE. North America, Central America, N. South America, Atlantic Ocean, W. Africa)

Entries with an * are at this distance for an extended period around this time.

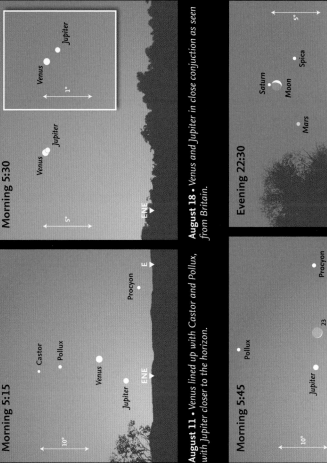

Morning 5:30

August 18 • Venus and Jupiter in close conjunction as seen from Britain.

Morning 5:15

August 11 • Venus lined up with Castor and Pollux, with Jupiter closer to the horizon.

Morning 5:45

August 23–24 • The Moon with Jupiter and Venus in the early morning.

Evening 22:30

August 31 • The Moon with Spica, Saturn and Mars. Seen from Britain there is no occultation of Saturn.

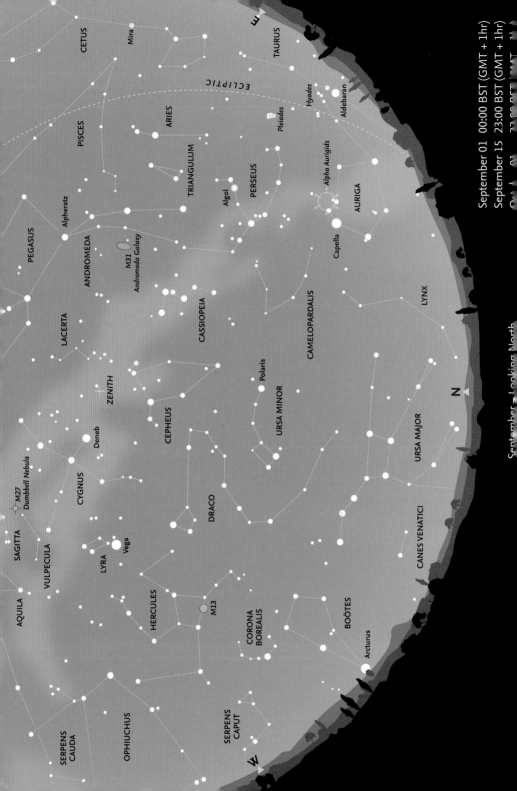

September – Looking North

September – Looking North

Ursa Major is now low in the north and to the north-west **Arcturus** and much of **Boötes** sink below the horizon later in the night and later in the month. In the north-east **Auriga** is beginning to climb higher in the sky. Later in the month, **Taurus**, with orange **Aldebaran** (α Tauri) and even **Gemini** with **Castor** and **Pollux** become visible in the east and north-east. Due east, **Andromeda** is now clearly visible, with the small constellations of **Triangulum** and **Aries** (the latter a zodiacal constellation) directly below it. Practically the whole of the Milky Way is visible, arching across the sky, both in the north and in the south. It is not particularly clear in Auriga, or even **Perseus**, but in **Cassiopeia** and on towards **Cygnus** the clouds of stars become easier to see. **Cepheus** is 'upside-down' near the zenith, and the head of **Draco** and **Hercules** beyond it are well placed for observation.

Meteors

Unfortunately, after the major Perseid shower in August, there is very little shower activity in September. There is one minor, but very extended, shower, known as the **Alpha Aurigids**, which actually has two peaks of activity. The first was on August 28 but the primary peak occurs on September 15. At either of the maxima, however, the hourly rate hardly reaches 10 meteors per hour, although the meteors are bright and relatively easy to photograph. Activity from this shower also extends into October. As a slight compensation for the lack of activity, however, in September the number of sporadic meteors observed reaches its highest rate at any time during the year.

Alpheratz (α Andromedae) forms the north-east corner of the Great Square of Pegasus. The constellation consists mainly of two lines of stars running towards Cassiopeia and Perseus.

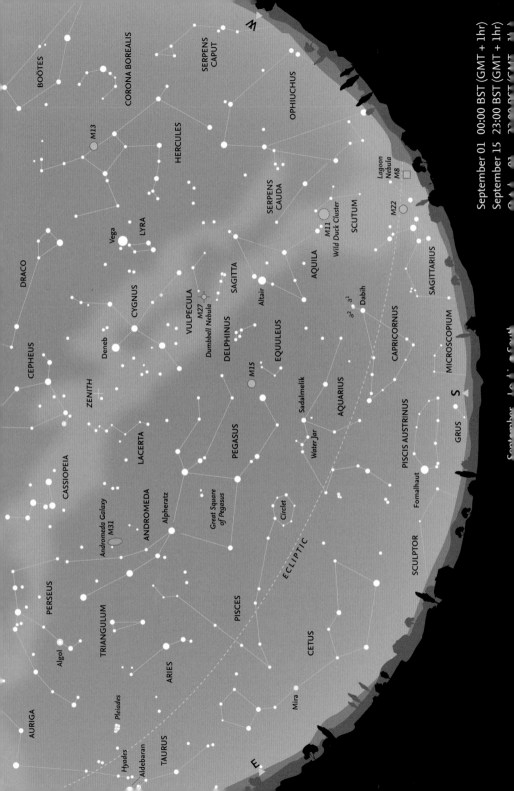

September 01 00:00 BST (GMT + 1hr)
September 15 23:00 BST (GMT + 1hr)

September – Looking South

The **Summer Triangle** is now high in the south-west, with the Great Square of **Pegasus** high in the south-east. Below Pegasus are the two zodiacal constellations of **Capricornus** and **Aquarius**. In what is otherwise an unremarkable constellation, α Capricorni is actually a visual binary, with the two stars (**Prima Giedi**, α¹ Cap, and **Secunda Giedi**, α² Cap) readily seen with the naked eye. **Dabih** (β Capricorni), just to the south, is also a double star, and the components are relatively easy to separate with binoculars. In Aquarius, just to the east of **Sadalmelik** (α Aquarii) there is a small asterism, consisting of four stars, resembling a tiny version of Cancer, known as the 'Water Jar'. Below Aquarius is a sparsely populated area of the sky, with just one bright star in the constellation of **Piscis Austrinus**. In classical illustrations, water is shown flowing from the 'Water Jar' towards bright **Fomalhaut** (α Piscis Austrini).

Another zodiacal constellation, **Pisces**, is now clearly visible to the east of Aquarius. Although faint, there is a distinctive asterism of stars, known as the 'Circlet' south of the Great Square and another line of faint stars to the east of Pegasus. Still farther down towards

the horizon is the constellation of **Cetus**, with the famous variable star, **Mira** (ο Ceti) at its centre. When Mira is at maximum brightness (around mag. 3.5) it is clearly visible to the naked eye, but it disappears as it fades towards minimum (about mag. 9.5 or less).

Pegasus is quite a large constellation, because apart from the Great Square, it consists of lines of stars stretching west towards the main band of the Milky Way.

The Moon's phases for September

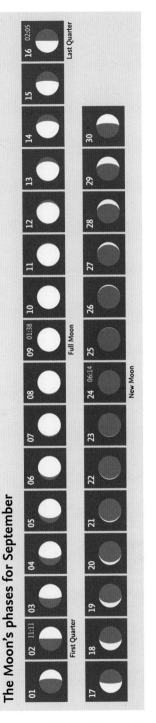

September – Moon and Planets

The Moon

Some days after Full Moon (September 9), on September 14–15, the waning gibbous Moon passes the *Pleiades* and *Aldebaran* shortly after midnight. Just before New Moon (September 24), the waning crescent is close to both *Jupiter* and *Regulus* in the early-morning sky on September 20–21. Later, after New Moon, over the three days September 27–29, the young waxing crescent passes close to *Saturn*, and (on September 29) both *Mars* and *Antares*.

The Planets

Although *Mercury* comes to greatest eastern elongation on September 21, it is so low on the western horizon that it remains invisible against the bright background shortly after sunset. Although it is in the dawn sky, *Venus* is bright throughout the month (mag. -3.9) so should be reasonably easy to detect shortly before sunrise. *Mars* begins the month in *Libra*, but ends it in *Scorpius*. On the evening of September 29 there is a chance to compare Mars (at mag. 0.8) with *Antares* – the 'Rival of Mars' – which is slightly fainter (at mag. 1.0). *Jupiter* is at magnitude -1.8 or -1.9 throughout the month. It continues direct motion out of *Cancer* and into *Leo*, appearing in the eastern sky before dawn. *Saturn* is at magnitude 0.6 throughout the month and may be glimpsed low in the western sky after sunset.

The path of the Sun and the planets along the ecliptic in September.

Calendar for September

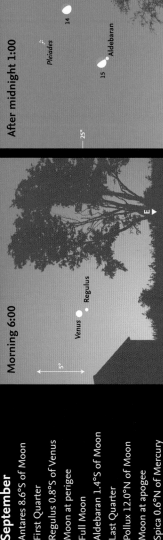

02	08:57	Antares 8.6°S of Moon
02	11:11	First Quarter
05	12:00*	Regulus 0.8°S of Venus
08	03:31	Moon at perigee
09	01:38	Full Moon
15	01:24	Aldebaran 1.4°S of Moon
16	02:05	Last Quarter
18	18:15	Pollux 12.0°N of Moon
20	14:22	Moon at apogee
21	02:00*	Spica 0.6°N of Mercury
21	19:16	Regulus 4.7°N of Moon
21	22:10	Mercury greatest elongation (26°E, mag. 0.0)
23	02:29	Autumnal equinox
24	06:14	New Moon
26	01:10	Spica 2.6°S of Moon
27	21:00*	Antares 3.1°S of Mars
28		Saturn occulted by Moon (E. Asia, Japan, NW Pacific, Hawaii)
29	14:37	Antares 8.7°S of Moon

Entries with an * are at this distance for an extended period around this time.

September 5 • Venus and Regulus in the early morning. The actual conjunction is in daylight (1:00 BST).

September 14–15 • The Moon passes the Pleiades and Aldebaran.

September 20–21 • The Moon passes Jupiter and Regulus.

September 27–29 • The Moon, Saturn, Mars and Antares. Mars is slightly brighter than its 'rival', Antares.

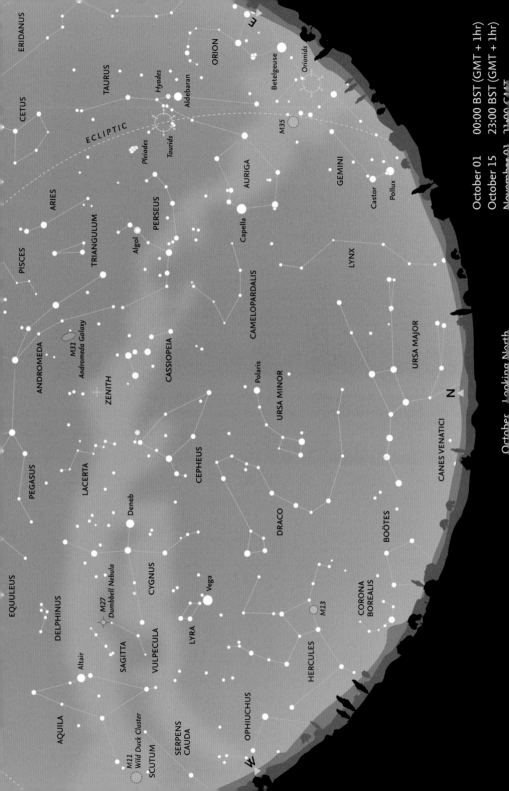

ERIDANUS

ORION

E

Orionids

Betelgeuse

Hyades

TAURUS

Aldebaran

CETUS

M35

ECLIPTIC

Taurids

GEMINI

AURIGA

Pleiades

ARIES

Castor

Pollux

PISCES

Algol

PERSEUS

Capella

TRIANGULUM

LYNX

ANDROMEDA

M31
Andromeda Galaxy

CAMELOPARDALIS

CASSIOPEIA

URSA MAJOR

ZENITH

Polaris

PEGASUS

URSA MINOR

N

LACERTA

CANES VENATICI

CEPHEUS

EQUULEUS

Deneb

DELPHINUS

DRACO

BOÖTES

M27
Dumbbell Nebula

CYGNUS

Altair

SAGITTA

VULPECULA

Vega

CORONA
BOREALIS

AQUILA

LYRA

M13

SERPENS
CAUDA

OPHIUCHUS

HERCULES

M11
Wild Duck Cluster

SCUTUM

W

October – Looking North

Ursa Major is grazing the horizon in the north, while high overhead are the constellations of *Cepheus*, *Cassiopeia* and *Perseus*, with the Milky Way between Cepheus and Cassiopeia at the zenith. *Auriga* is now clearly visible in the east, as is *Taurus* with the *Pleiades*, *Hyades* and orange *Aldebaran*. Also in the east, *Orion* and *Gemini* are starting to rise clear of the horizon.

The constellations of *Boötes* and *Corona Borealis* are now essentially lost to view in the north-west, and *Hercules* is also descending towards the western horizon. The three stars of the Summer Triangle are still clearly visible, although *Aquila* and *Altair* are beginning to approach the horizon in the west. On October 26 Summer Time ends in Europe with Britain reverting to Greenwich Mean Time, and Europe to Central European Time.

Meteors

The *Orionids* are the major (and fairly reliable) meteor shower active in October. As with the *Eta Aquarid* shower in May, the Orionids are associated with Comet P/Halley. During this second of the yearly passes through the stream of particles from the comet, slightly fewer meteors are seen than in the Eta Aquarids, but conditions are more favourable for northern observers. In both showers the meteors are very fast, and many leave persistent trains. Although the Orionid maximum is quoted as being October 21, in fact there is a very broad maximum, lasting about a week from October 20 to 27, with hourly rates between 15 and 30.

Towards the end of the month, another shower (the *Taurids*) begins to show activity, which peaks early in November.

The constellation of Cassiopeia is a familiar sight among the northern circumpolar constellations. It is always above the horizon, on the opposite side of Polaris (the North Star) to the equally well-known asterism of the seven stars of the Plough.

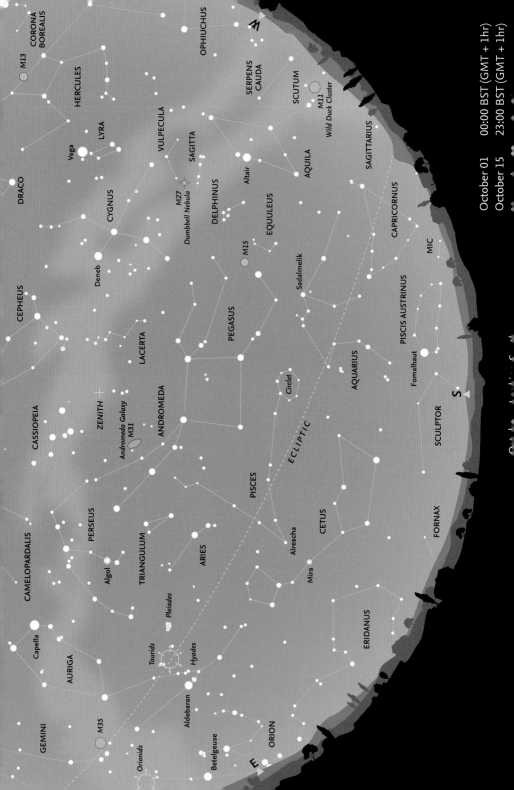

October 01   00:00 BST (GMT + 1hr)
October 15   23:00 BST (GMT + 1hr)

October – Looking South

The Great Square of *Pegasus* dominates the southern sky, framed by the two chains of stars that form the constellation of *Pisces*, together with *Alrescha* (α Piscium) at the point where the two lines of stars join. Also clearly visible is the constellation of *Cetus*, below Pegasus and Pisces. Although *Capricornus* is beginning to disappear, *Aquarius* to its west is well placed in the south, with solitary *Fomalhaut* and the constellation of *Piscis Austrinus* beneath it, close to the horizon.

The main band of the Milky Way and the Great Rift runs down from *Cygnus*, through *Vulpecula*, *Sagitta* and *Aquila* towards the western horizon. *Delphinus* and the tiny, unremarkable constellation of *Equuleus* lie between the band of the Milky Way and Pegasus. *Andromeda* is clearly visible high in the sky to the south-east, with the small constellation of *Triangulum* and the zodiacal constellation of *Aries* below it. *Perseus* is high in the east, and by now the *Pleiades* and *Taurus* are well clear of the horizon. Later in the night, and later in the month, *Orion* rises in the east, a sign that the autumn season has arrived and of the steady approach of winter.

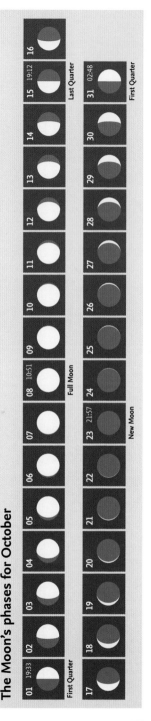

The constellation of Aquarius is one of the zodiacal constellations that is visible in late summer and early autumn. The four stars forming the 'Y'-shape of the 'Water Jar' may be seen to the east of Sadalmelik, the brightest star (top centre).

The Moon's phases for October

01 19:33	02	03	04	05	06	07	08 10:51
First Quarter							Full Moon

09	10	11	12	13	14	15 19:12	16
						Last Quarter	

17	18	19	20	21	22	23 21:57	24
							New Moon

25	26	27	28	29	30	31 02:48
						First Quarter

October – Moon and Planets

The Moon

On October 8 there is a total lunar eclipse, unfortunately not visible from Europe, and seen only from western North America and across the Pacific. On October 12, the waning gibbous Moon passes close to *Aldebaran* in the south-western morning sky. A few days later, on October 18–19, it may be seen on the opposite side of the sky, in the east, near *Jupiter* and *Regulus* just before dawn. At New Moon on October 23 there is a partial solar eclipse. This eclipse, too, is visible only from western North America. Two days later, on October 25, the Moon is extremely close to *Saturn* and actually occults the planet for observers in North America and the Iberian Peninsula. On the three following days (October 26–28), the waxing crescent Moon passes *Antares* and *Mars*, low on the south-western horizon after sunset.

The Planets

Late in the month, *Mercury* becomes visible in the morning sky, rapidly brightening from mag. 2.3 on October 22 as it moves towards greatest western elongation (on November 1). *Venus* is too close to the Sun for observation and it comes to superior conjunction (on the far side of the Sun) on October 25. *Mars* (moving from *Scorpius* into *Sagittarius* over the course of the month and almost constant at mag. 0.8–0.9) is low in the south-west after sunset. *Jupiter* is roughly half-way between *Cancer* and *Leo*, in the morning sky, brightening slightly from mag. -1.9 at the beginning of the month to mag. -2.1 by the end. *Saturn* is slowly creeping eastwards, still in *Libra* but difficult to see in the evening twilight. It is at mag. 0.6 for most of the month, but brightens very slightly to mag. 0.5 in the last few days.

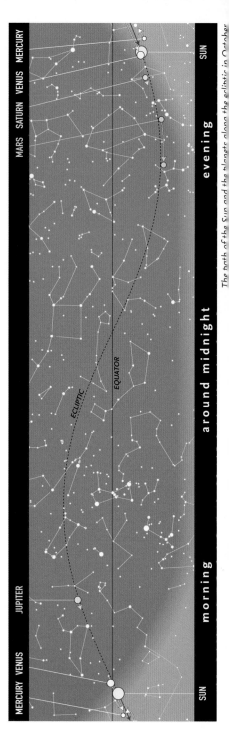

The path of the Sun and the planets along the ecliptic in October.

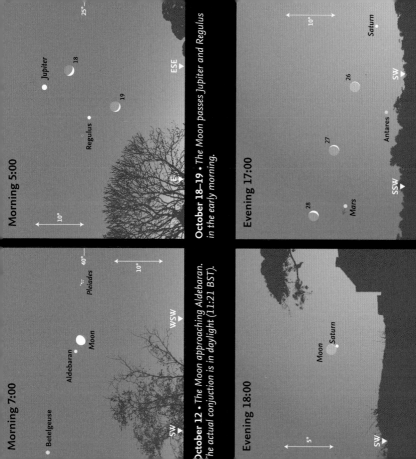

Morning 7:00

Betelgeuse

Aldebaran

Moon

Pleiades

40°

10°

WSW

SW

October 12 • The Moon approaching Aldebaran. The actual conjuction is in daylight (11:21 BST).

Morning 5:00

25°

18

Jupiter

19

Regulus

10°

E

ESE

October 18–19 • The Moon passes Jupiter and Regulus in the early morning.

Evening 18:00

Moon

Saturn

5°

SW

October 25 • The Moon and Saturn. Seen from Britain, Saturn is not occulted by the Moon.

Evening 17:00

10°

28

27

26

Mars

Antares

Saturn

SSW

SW

October 26–28 • The Moon continues and passes Antares and Mars. The fauve is for 17:00 GMT.

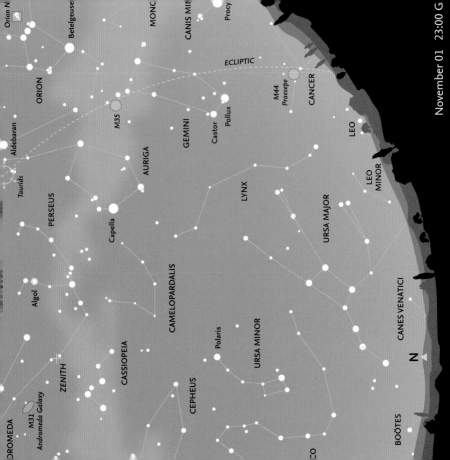

November – Looking North

Aquila has now disappeared below the horizon, but two of the stars of the Summer Triangle, **Vega** in **Lyra** and **Deneb** in **Cygnus**, are still visible in the west. The head of **Draco** is now low in the north-west and only a small portion of **Hercules** remains above the horizon. The Milky Way arches overhead, with the denser star clouds in the west and the less heavily populated region through **Auriga** and **Monoceros** in the east. High overhead, **Cassiopeia** is near the zenith and **Cepheus** has swung round to the north-west, while Auriga is now high in the north-east. **Gemini**, with **Castor** and **Pollux** is well clear of the eastern horizon, and even **Procyon** (α Canis Minoris) is just climbing into view almost due east.

Meteors

The **Taurid** shower, which began in mid-October, reaches maximum – although with only a moderate rate of about 10 meteors per hour – on November 3, but gradually trails off, ending around November 25. Far more striking, however, are the **Leonids**, which have a short period of activity (November 15–20), with maximum on November 17. This shower is associated with Comet P/Tempel-Tuttle and has shown extraordinary activity on various occasions with many thousands of meteors per hour. High rates were seen in 1999, 2001 and 2002 (reaching about 3,000 meteors per hour) but have fallen dramatically since then, down to about 60 per hour in 2012. The meteors are very fast – the fastest shower meteors recorded, about 70 km per second – and often leave trails that are very persistent. The shower is very rich in faint meteors. In 2014, maximum is a few days before New Moon (November 22), so the waning crescent should not create too much interference. Observing conditions are always best after midnight. A radiant map for the Leonids can be found on page 16.

The constellation of Auriga, with Capella, its brightest star, and the distinctive triangle of the 'Kids' to its west (centre right) are prominent features in the northern sky. Capella itself is circumpolar for latitude 50° north and skims the horizon during the summer, but never actually disappears below it.

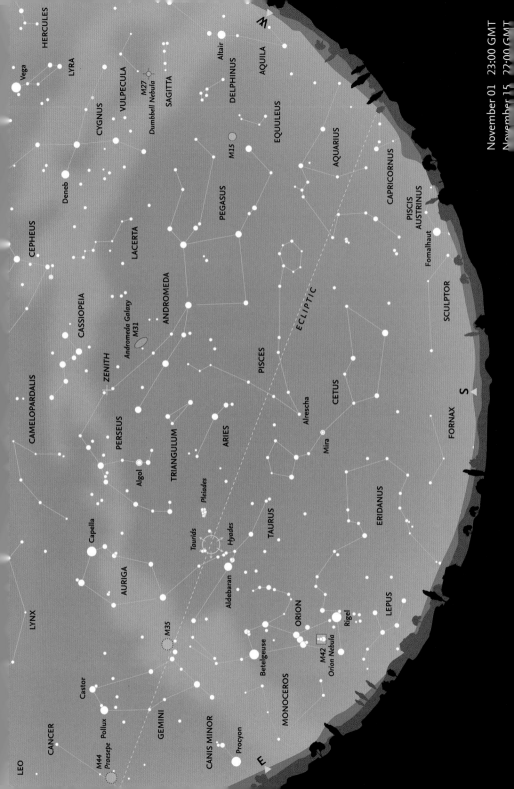

November – Looking South

Orion has now risen above the eastern horizon, and part of the long, straggling constellation of **Eridanus** (which begins near **Rigel**) is visible to the west of Orion. Higher in the sky, **Taurus**, with the **Pleiades** open cluster and orange **Aldebaran** are now easy to observe. To their west, both **Pisces** and **Cetus** are close to the meridian. In the south-west, **Capricornus** has slipped below the horizon, but **Aquarius** remains visible. Even farther west, **Altair** may be seen early in the night, but most of **Aquila** has already disappeared from view. **Delphinus**, together with **Sagitta** and **Vulpecula** in the Milky Way will soon vanish for another year. Both **Pegasus** and **Andromeda** are easy to see, and one of the lines of stars that make up Andromeda finishes close to the zenith, which is also close to one of the outlying stars of **Perseus**, high in the east.

Traditional representations of Pisces show two fish, tied together with ribbons. The star at the knot is Alrescha (α Piscium), which together with the two lines of stars, forms a broad arrow point down towards the variable star Mira in Cetus.

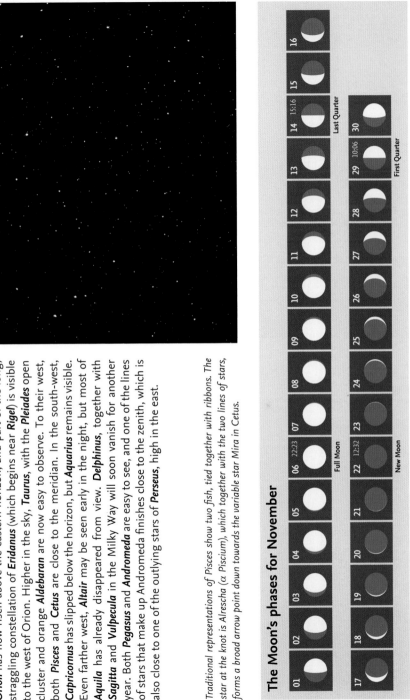

The Moon's phases for November

01	02	03	04	05	06 22:23
07	08	09	10	11	12
13	14 15:16	15	16		

Full Moon (06)

Last Quarter (14)

17	18	19	20	21	22 12:32
23	24	25	26	27	28
29 10:06	30				

New Moon (22)

First Quarter (29)

November – Moon and Planets

The Moon

As in October, the Moon passes close to the *Pleiades* and *Aldebaran* in *Taurus*, but this time (on November 7–8), it is just after Full Moon (November 6) and the events take place almost due east. Once again, too, on November 14 and 15, the Moon passes *Jupiter* and *Regulus* in the eastern sky. A few days later (November 19–21) the waning crescent Moon passes *Spica* and *Mercury*, just before dawn. Mercury, although at mag. -0.8, however, is now very low down on the horizon.

The Planets

Mercury reaches greatest western elongation, 19° west of the Sun and at mag. -0.6 on November 1. (November apparitions are particularly favourable because of the high angle the orbit makes with the Earth's horizon.) *Venus* is too close to the Sun to be observed, even though it is bright (mag. -4.0 to -3.9). *Mars* is in *Sagittarius* and very low in the evening sky, at mag. 0.9–1.0. *Jupiter* is moving slowly into *Leo*, very gradually brightening from mag. -2.1 to mag. -2.2. *Saturn* remains in *Libra*, with direct motion, and is at mag. 0.5 throughout the month.

The path of the Sun and the planets along the ecliptic in November.

Calendar for November

01	12:39	Mercury greatest elongation (19°W, mag. -0.6)
02		Daylight Saving Time ends (USA)
03		Taurid meteor maximum
03	00:29	Moon at perigee
06	22:23	Full Moon
08	20:04	Aldebaran 1.4°S of Moon
12	09:59	Pollux 12.1°N of Moon
13	09:00*	Saturn 1.6°N of Venus
14	15:16	Last Quarter
14	17:45	Jupiter 5.3°N of Moon
15	01:56	Moon at apogee
15	10:29	Regulus 4.6°N of Moon
15–20		Leonid meteor shower
17		Leonid meteor maximum
18	08:50	Saturn in conjunction with Sun
19	16:24	Spica 2.6°S of Moon
21	17:15	Mercury 1.9°S of Moon
22	05:22	Saturn 1.3°S of Moon
22	12:32	New Moon
23	01:03	Venus 3.9°S of Moon
23	03:36	Antares 8.6°S of Moon
24	04:00*	Antares 4.6°S of Venus
26	09:00*	Mercury 1.7°S of Saturn
26	10:08	Mars 6.6°S of Moon
27	23:12	Moon at perigee
29	10:06	First Quarter

*entries with an * are at this distance for an extended*

Before sunrise

Mercury

Nov. 01
Oct. 27 — Nov. 06
Nov. 11
Oct. 22 — Nov. 16
Nov. 21
Nov. 26

10°

Morning apparition of Mercury · *Positions are shown in relation to the sun.*

Evening 20:00

Pleiades

07

08 Aldebaran

10°

ENE   E

November 7–8 · *The Moon passes the Pleiades and Aldebaran in the evening sky.*

Morning 4:00

14

Jupiter

Regulus

15

10°

40°

ESE   SE

November 14–15 · *The Moon with Jupiter and*

Morning 7:00

19   Spica

20

21   Mercury

10°

ESE   SE

November 19–21 · *The crescent Moon passes Spica and*

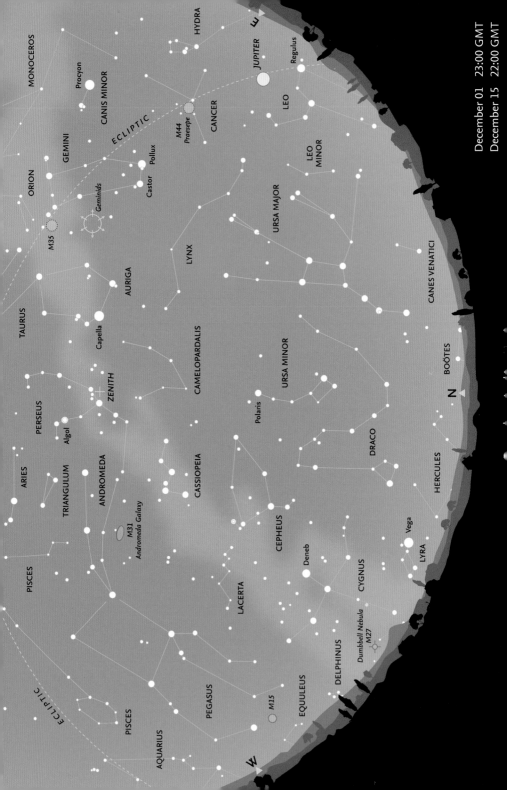

December – Looking North

Ursa Major has now swung round and is starting to 'climb' in the east. The fainter stars in the southern part of the constellation are now fully in view. The other bear, *Ursa Minor*, 'hangs' below *Polaris* in the north. Directly above it is the faint constellation of *Camelopardalis*, with the other inconspicuous circumpolar constellation, *Lynx*, to its east. *Vega* (α Lyrae) is skimming the horizon in the north-west, but *Deneb* (α Cygni) and most of *Cygnus* remain visible farther west. In the east, *Regulus* (α Leonis) and the constellation of *Leo* are beginning to rise above the horizon. *Cancer* stands high in the east, with *Gemini* even higher in the sky. *Perseus* is at the zenith, with *Auriga* and *Capella* between it and Gemini. Because it is so high in the sky, now is a good time to examine the star clouds of the fainter portion of the Milky Way, between *Cassiopeia* in the west to Gemini and *Orion* in the east.

Meteors

There is one significant meteor shower in December (the last major shower of the year). This is the *Geminid* shower, which is visible over the period December 7–15 and comes to maximum on December 13. It is one of the most active showers of the year, and indeed in some years is the most active, with a peak rate of around 100 meteors per hour. For many years it posed a problem, because the meteors were found to have a much higher density than other meteors (which are derived from cometary material). It was eventually established that the Geminids and the asteroid Phaeton had similar orbits. So the Geminids are assumed to consist of denser, rocky material. They are slower than most other meteors and often appear to last longer. The brightest often break up into numerous luminous fragments that follow similar paths across the sky.

Although the seven stars that form the asterism of the Plough are familiar and readily recognized by virtually everyone, Ursa Major (the Great Bear) is one of the largest constellations and the fainter stars to the east and south are often overlooked.

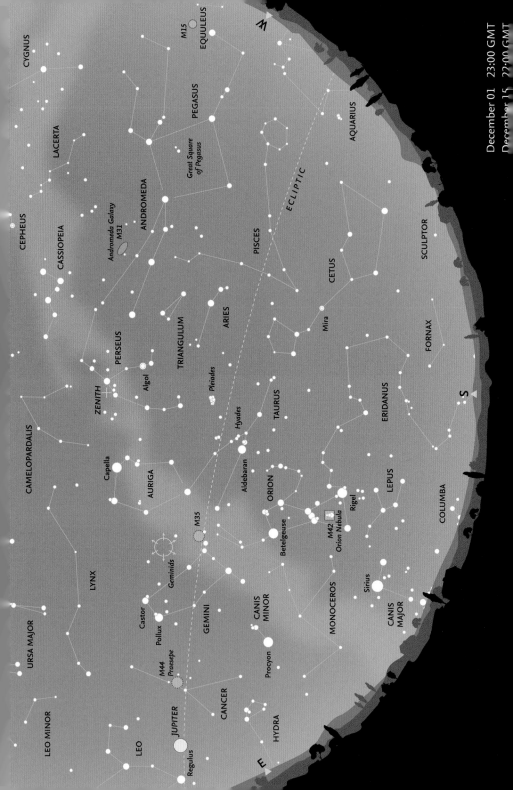

December 01 23:00 GMT
December 15 22:00 GMT

December – Looking South

The fine open cluster of the *Pleiades* is due south around 22:00, with the *Hyades* open cluster, *Aldebaran* and the rest of *Taurus* clearly visible to the east. *Auriga* (with *Capella*) and *Gemini* (with *Castor* and *Pollux*) are both well-placed for observation. *Orion* has made a welcome return to the winter sky, and both *Canis Minor* (with *Procyon*) and *Canis Major* (with *Sirius*, the brightest star in the sky) are now well above the horizon. The small, poorly known constellation of *Lepus* lies to the south of Orion. In the east, *Aquarius* has now disappeared, and *Cetus* is becoming lower, but *Pisces* is still easily seen, as are the constellations of *Aries*, *Triangulum* and *Andromeda* above it. The Great Square of *Pegasus* is starting to plunge down towards the eastern horizon, and because of its orientation on the sky appears more like a large diamond, standing on one point, than a square.

The constellation of Taurus contains two contrasting open clusters: the compact Pleiades, with its striking blue-white stars, and the more scattered, but much closer, 'V'-shaped Hyades. Orange Aldebaran (α Tauri) is not related to the Hyades, but lies between it and the Earth.

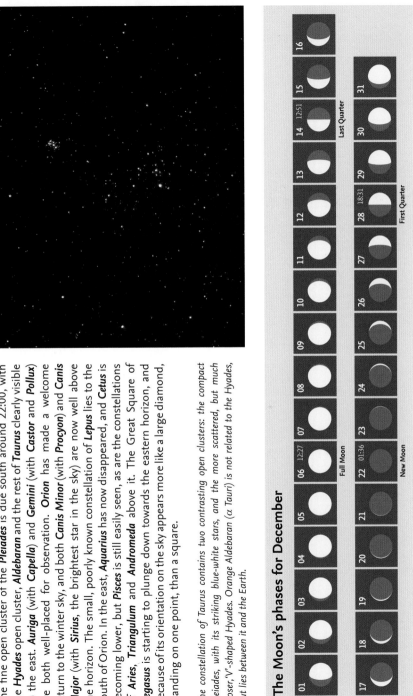

The Moon's phases for December

01	02	03	04	05	06 12:27
07	08	09	10	11	12
13	14 12:51	15	16		

Full Moon

Last Quarter

17	18	19	20	21	22 01:36
23	24	25	26	27	28 18:31
29	30	31			

New Moon

First Quarter

December – Moon and Planets

The Moon

On the day of Full Moon (December 6) the Moon passes close to *Aldebaran* in the western morning sky. A few days later, on December 12, the waning gibbous Moon appears below *Jupiter* and *Regulus* in the morning sky to the south-west. A few days later again, on December 17–20, the waning crescent glides past *Spica* and *Saturn* in *Libra* just before dawn. On December 23–25, the days immediately after New Moon (December 22), the narrow crescent Moon appears in the western sky after sunset together with *Venus* (very low on the horizon, but at mag. -3.9) and *Mars* (mag. 1.1).

The Planets

Mercury is too close to the Sun to be visible this month. It reaches superior conjunction (beyond the Sun) on December 8. Although *Venus* is bright (mag. -3.9) throughout the month, it is also very close to the Sun, but may be glimpsed after sunset in the western sky. *Mars* (mag. 1.0–1.1) is in *Capricornus*, and may be visible for a short period in the western sky shortly after sunset. *Jupiter* is in *Leo* and remains fairly bright (mag. -2.2 to -2.4) throughout the month. It begins retrograde motion on December 14, which continues to the end of the year. (It will reach opposition on 6 February 2015.) *Saturn* has remained in *Libra* for the whole year, but ends 2014 on the border of *Scorpius*. It fades very slightly from mag. 0.5 to 0.6 during December.

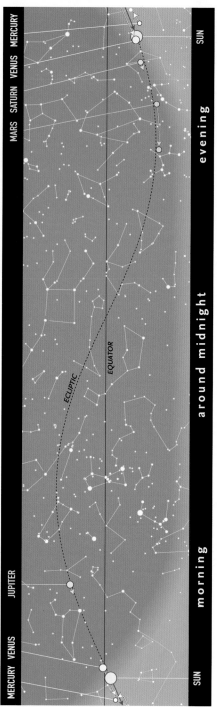

The path of the Sun and the planets along the ecliptic in December.

Calendar for December

04	01:00*	Antares 4.0°S of Mercury
06	04:58	Aldebaran 1.5°S of Moon
06	12:27	Full Moon
07–15		Geminid meteor shower
08	09:51	Mercury superior conjunction
09	18:31	Pollux 11.9°N of Moon
12	03:43	Jupiter 5.1°N of Moon
12	18:31	Regulus 4.7°N of Moon
12	23:03	Moon at apogee
13		Geminid meteor maximum
14	12:51	Last Quarter
17	01:28	Spica 2.8°S of Moon
19	20:36	Saturn 1.5°S of Moon
20	13:05	Antares 8.7°S of Moon
21	23:03	Winter solstice
22	01:36	New Moon
22	16:50	Mercury 7.0°S of Moon
23	04:36	Venus 6.2°S of Moon
24	16:42	Moon at perigee
25	07:34	Mars 5.7°S of Moon
28	18:31	First Quarter

*Entries with an * are at this distance for an extended period around this time.*

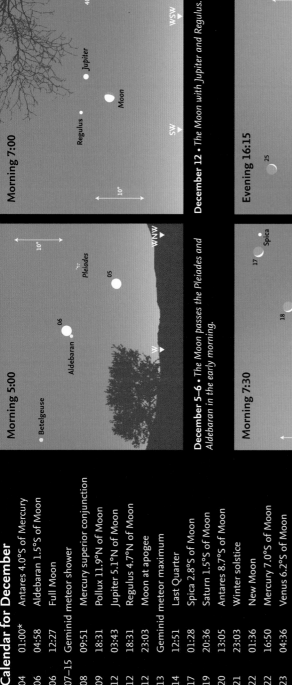

Morning 5:00

December 5–6 • *The Moon passes the Pleiades and Aldebaran in the early morning.*

Morning 7:00

December 12 • *The Moon with Jupiter and Regulus.*

Morning 7:30

December 17–20 • *The Moon passes Spica, Saturn and Antares. It will not be easy to see Antares in the twilight.*

Evening 16:15

December 23–25 • *The Moon with Venus and Mars shortly after sunset.*

Glossary and Tables

aphelion	The point on an orbit that is farthest from the Sun.
apogee	The point on its orbit at which the Moon is farthest from the Earth.
appulse	The apparently close approach of two celestial objects; two planets, or a planet and star.
astronomical unit	(AU) The mean distance of the Earth from the Sun, 149,597,870 km.
celestial equator	The great circle on the celestial sphere that is in the same plane as the Earth's equator.
celestial sphere	The apparent sphere surrounding the Earth on which all celestial bodies (stars, planets, etc.) seem to be located.
conjunction	The point in time when two celestial objects have the same celestial longitude. In the case of the Sun and a planet, superior conjunction occurs when the planet lies on the far side of the Sun (as seen from Earth). For Mercury and Venus, inferior conjuction occurs when they pass between the Sun and the Earth.
direct motion	Motion from west to east on the sky.
ecliptic	The apparent path of the Sun across the sky throughout the year. Also the plane of the Earth's orbit in space.
elongation	The point at which an inferior planet has the greatest angular distance from the Sun, as seen from Earth.
equinox	The two points during the year when night and day have equal duration. Also: the points on the sky at which the ecliptic intersects the celestial equator. The vernal equinox is of particular importance in astronomy.
gibbous	The stage in the sequence of phases at which the illumination of a body lies between half and full. In the case of the Moon, the term is applied to phases between First Quarter and Full, and between Full and Last Quarter.
inferior planet	Either of the planets Mercury or Venus, which have orbits inside that of the Earth.
magnitude	The brightness of a star, planet, or other celestial body. It is a logarithmic scale, where larger numbers indicate fainter brightness. A difference of 5 in magnitude indicates a difference of 100 in actual brightness, thus a first-magnitude star is 100 times as bright as one of sixth magnitude.
meridian	The great circle passing through the north and south poles of a body and the observer's position; or the corresponding great circle on the celestial sphere that passes through the North and South Celestial Poles and also through the observer's zenith.
nadir	The point on the celestial sphere directly beneath the observer's feet, opposite the zenith.
occultation	The disappearance of one celestial body behind another, such as when stars or planets are hidden behind the Moon.
opposition	The point on a superior planet's orbit at which it is directly opposite the Sun in the sky.
perigee	The point on its orbit at which the Moon is closest to the Earth.
perihelion	The point on an orbit that is closest to the Sun.
retrograde motion	Motion from east to west on the sky.
superior planet	A planet that has an orbit outside that of the Earth.
vernal equinox	The point at which the Sun, in its apparent motion along the ecliptic, crosses the celestial equator from south to north. Also known as the First Point of Aries.
zenith	The point directly above the observer's head.
zodiac	A band, 8° on either side of the ecliptic, within which the Moon and planets appear to move. It consists of 12 equal areas, originally named after the constellation that once lay within it.

The Greek Alphabet

α	Alpha	ε	Epsilon	ι	Iota	ν	Nu	ρ	Rho	φ (φ)	Phi
β	Beta	ζ	Zeta	κ	Kappa	ξ	Xi	σ (ς)	Sigma	χ	Chi
γ	Gamma	η	Eta	λ	Lambda	ο	Omicron	τ	Tau	ψ	Psi
δ	Delta	θ (ϑ)	Theta	μ	Mu	π	Pi	υ	Upsilon	ω	Omega

The Constellations

There are 88 constellations covering the whole of the celestial sphere, but 24 of these in the southern hemisphere can never be seen (even in part) from the latitude of Britain and Ireland, so are omitted from this table. The names themselves are expressed in Latin, and the names of stars are frequently given by Greek letters followed by the genitive of the constellation name. The genitives and English names of the various constellations are included.

Name	Genitive	Abbr.	English name
Andromeda	Andromeda	And	Andromeda
Antlia	Antliae	Ant	Air Pump
Aquarius	Aquarii	Aqr	Water Bearer
Aquila	Aquilae	Aql	Eagle
Aries	Arietis	Ari	Ram
Auriga	Aurigae	Aur	Charioteer
Boötes	Boötis	Boo	Herdsman
Camelopardalis	Camelopardalis	Cam	Giraffe
Cancer	Cancri	Cnc	Crab
Canes Venatici	Canum Venaticorum	CVn	Hunting Dogs
Canis Major	Canis Majoris	CMa	Big Dog
Canis Minor	Canis Minoris	CMi	Little Dog
Capricornus	Capricorni	Cap	Sea Goat
Cassiopeia	Cassiopeiae	Cas	Cassiopeia
Centaurus	Centauri	Cen	Centaur
Cepheus	Cephei	Cep	Cepheus
Cetus	Ceti	Cet	Whale
Columba	Columbae	Col	Dove
Coma Berenices	Coma Berenicis	Com	Berenice's Hair
Corona Australis	Coronae Australis	CrA	Southern Crown
Corona Borealis	Coronae Borealis	CrB	Northern Crown
Corvus	Corvi	Crv	Crow
Crater	Crateris	Crt	Cup
Cygnus	Cygni	Cyg	Swan
Delphinus	Delphini	Del	Dolphin
Draco	Draconis	Dra	Dragon
Equuleus	Equulei	Equ	Little Horse
Eridanus	Eridani	Eri	River Eridanus
Fornax	Fornacis	For	Furnace
Gemini	Geminorum	Gem	Twins
Hercules	Herculis	Her	Hercules
Hydra	Hydrae	Hya	Water Snake

Name	Genitive	Abbr.	English name
Lacerta	Lacertae	Lac	Lizard
Leo	Leonis	Leo	Lion
Leo Minor	Leonis Minoris	LMi	Little Lion
Lepus	Leporis	Lep	Hare
Libra	Librae	Lib	Scales
Lupus	Lupi	Lup	Wolf
Lynx	Lyncis	Lyn	Lynx
Lyra	Lyrae	Lyr	Lyre
Microscopium	Microscopii	Mic	Microscope
Monoceros	Monocerotis	Mon	Unicorn
Ophiuchus	Ophiuchi	Oph	Serpent Bearer
Orion	Orionis	Ori	Orion
Pegasus	Pegasi	Peg	Pegasus
Perseus	Persei	Per	Perseus
Pisces	Piscium	Psc	Fishes
Piscis Austrinus	Piscis Austrini	PsA	Southern Fish
Puppis	Puppis	Pup	Stern
Pyxis	Pyxidis	Pyx	Compass
Sagitta	Sagittae	Sge	Arrow
Sagittarius	Sagittarii	Sgr	Archer
Scorpius	Scorpii	Sco	Scorpion
Sculptor	Sculptoris	Scl	Sculptor
Scutum	Scuti	Sct	Shield
Serpens	Serpentis	Ser	Serpent
Sextans	Sextantis	Sex	Sextant
Taurus	Tauri	Tau	Bull
Triangulum	Trianguli	Tri	Triangle
Ursa Major	Ursae Majoris	UMa	Great Bear
Ursa Minor	Ursae Minoris	UMi	Lesser Bear
Vela	Velorum	Vel	Sail
Virgo	Virginis	Vir	Virgin
Vulpecula	Vulpeculae	Vul	Fox

Some common asterisms

Belt of Orion	δ, ε, and ζ Orionis
Big Dipper	α, β, γ, δ, ε, ζ, and η Ursae Majoris
Circlet	γ, θ, ι, λ, and κ Piscium
Guards (or Guardians)	β and γ Ursae Minoris
Head of Cetus	α, γ, ξ², μ, and λ Ceti
Head of Draco	β, γ, ξ, and ν Draconis
Head of Hydra	δ, ε, ζ, η, ρ, and σ Hydrae
Keystone	ε, ζ, η, and π Herculis
Kids	ε, ζ, and η Aurigae
Little Dipper	β, γ, η, ζ, ε, δ, and α Ursae Minoris
Lozenge	= Head of Draco
Milk Dipper	ζ, γ, σ, φ, and λ Sagittarii
Plough	α, β, γ, δ, ε, ζ, and η Ursae Majoris
Pointers	α and β Ursae Majoris
Sickle	α, η, γ, ζ, μ, and ε Leonis
Square of Pegasus	α, β, and γ Pegasi with α Andromedae
Sword of Orion	θ and ι Orionis
Teapot	γ, ε, δ, λ, φ, σ, τ, and ζ Sagittarii
Wain (or Charles' Wain)	= Plough
Water Jar	γ, η, κ, and ζ Aquarii
Y of Aquarius	= Water Jar

Acknowledgements

Dennis Buczynski, Portmahomack, Ross-Shire: p. 51 (NLC image)
Storm Dunlop, Chichester: p. 21 (Orion)
Steve Edberg, La Cañada, California: all other constellation photographs

Editorial support was provided by Colin Stuart, Astronomer at Royal Observatory Greenwich, part of Royal Museums Greenwich.

Further Information

Books

Bone, Neil (1993), *Observer's Handbook: Meteors*, George Philip, London & Sky Publ. Corp., Cambridge, Mass.

Cook, J., ed. (1999), *The Hatfield Photographic Lunar Atlas*, Springer-Verlag, New York

Dunlop, Storm (2012), *Practical Astronomy*, 3rd edn, Philip's, London

Dunlop, Storm (1999), *Wild Guide to the Night Sky*, HarperCollins, London

Dunlop, Storm, Rükl, Antonin & Tirion, Wil (2005), *Collins Atlas of the Night Sky*, HarperCollins, London

Ridpath, Ian, ed. (1998), *Norton's Star Atlas*, 19th edn, Longman, London

Ridpath, Ian, ed. (2003), *Oxford Dictionary of Astronomy*, 2nd edn, Oxford University Press, Oxford

Ridpath, Ian & Tirion, Wil (2012), *Monthly Sky Guide*, 9th edn, Cambridge Universtiy Press

Ridpath, Ian & Tirion, Wil (2011), *Collins Pocket Guide Stars and Planets*, 4th edn, HarperCollins, London

Ridpath, Ian & Tirion, Wil (2004), *Collins Gem – Stars*, HarperCollins, London

Rükl, Antonín (1990), *Hamlyn Atlas of the Moon*, Hamlyn, London & Astro Media Inc., Milwaukee

Rükl, Antonín (2004), *Atlas of the Moon*, Sky Publishing Corp., Cambridge, Mass.

Scagell, Robin (2000), *Philip's Stargazing with a Telescope*, George Philip, London

Tirion, Wil (2011), *Cambridge Star Atlas*, 4th edn, Cambridge University Press, Cambridge

Tirion, Wil & Sinnott, Roger (1999), *Sky Atlas 2000.0*, 2nd edn, Sky Publishing Corp., Cambridge, Mass. & Cambridge University Press, Cambridge

Journals

Astronomy, Astro Media Corp., 21027 Crossroads Circle, P.O. Box 1612, Waukesha, WI 53187-1612 USA. http://www.astronomy.com

Astronomy Now, Pole Star Publications, PO Box 175, Tonbridge, Kent TN10 4QX UK. http://www.astronomynow.com

Sky & Telescope, Sky Publishing Corp., Cambridge, MA 02138-1200, USA. http://www.skyandtelescope.com/

Societies

British Astronomical Association, Burlington House, Piccadilly, London W1J 0DU. http://www.britastro.org/
The principal British organization for amateur astronomers (with some professional members), particularly for those interested in carrying out observational programmes. Its membership is, however, worldwide. It publishes fully refereed, scientific papers and other material in its well-regarded Journal.

Federation of Astronomical Societies, Secretary: Ken Sheldon, Whitehaven, Maytree Road, Lower Moor, Pershore, Worcs. WR10 2NY. http://www.fedastro.org.uk/fas/
An organization that is able to provide contact information for local astronomical societies in the United Kingdom.

Royal Astronomical Society, Burlington House, Piccadilly, London W1J 0BQ. http://www.ras.org.uk/
The premier astronomical society, with membership primarily drawn from professionals and experienced amateurs. It has an exceptional library and is a designated centre for the retention of certain classes of astronomical data. Its publications are the standard medium for dissemination of astronomical research.

Society for Popular Astronomy, 36 Fairway, Keyworth, Nottingham NG12 5DU.
http://www.popastro.com/
A society for astronomical beginners of all ages, which concentrates on increasing members understanding and enjoyment, but which does have some observational programmes. Its journal is entitled *Popular Astronomy*.

Software

Planetary, Stellar and Lunar Visibility, (Planetary and eclipse freeware): Alcyone Software, Germany.
http://www.alcyone.de
Redshift, Redshift-Live. http://www.redshift-live.com/en/
Starry Night & Starry Night Pro, Sienna Software Inc., Toronto, Canada. http://www.starrynight.com

Internet sources

There are numerous sites with information about all aspects of astronomy, and all of those have numerous links. Although many amateur sites are excellent, treat any statements and data with caution. The sites listed below offer accurate information. Please note that the URLs may change. If so, use a good search engine, such as Google, to locate the information source.

Information

Auroral information Michigan Tech: http://www.geo.mtu.edu/weather/aurora/
Comets JPL Solar System Dynamics: http://ssd.jpl.nasa.gov/
Meteor Showers Online Gary Kronk's Home Page: http://meteorshowersonline.com
Deep-sky objects Saguaro Astronomy Club Database: http://www.virtualcolony.com/sac/
Eclipses: NASA Eclipse Page: http://eclipse.gsfc.nasa.gov/eclipse.html
Moon (inc. Atlas) Inconstant Moon: http://www.inconstantmoon.com/
Planets Planetary Fact Sheets: http://nssdc.gsfc.nasa.gov/planetary/planetfact.html
Satellites (inc. International Space Station)
        Heavens Above: http://www.heavens-above.com/
        Visual Satellite Observer's: http://www.satobs.org/
Star Chart National Geographic
        Chart: http://www.nationalgeographic.com/features/97/stars/chart/index.html
What's Visible Skyhound: http://www.skyhound.com/sh/skyhound.html
        Skyview Cafe: http://www.skyviewcafe.com
        Stargazer: http://www.outerbody.com/stargazer/ (choose 'File, New View', click & drag to change)

Institutes and Organizations

European Space Agency: http://www.esa.int/
International Dark-Sky Association: http://www.darksky.org/
Jet Propulsion Laboratory: http://www.jpl.nasa.gov/
Lunar and Planetary Institute: http://www.lpi.usra.edu/
National Aeronautics and Space Administration: http://www.hq.nasa.gov/
Solar Data Analysis Center: http://umbra.gsfc.nasa.gov/
Space Telescope Science Institute: http://www.stsci.edu/public.html